Applications of Operational Research and Mathematical Models in Management

Applications of Operational Research and Mathematical Models in Management

Editor

Miltiadis Chalikias

MDPI • Basel • Beijing • Wuhan • Barcelona • Belgrade • Manchester • Tokyo • Cluj • Tianjin

Editor
Miltiadis Chalikias
University of West Attica
Greece

Editorial Office
MDPI
St. Alban-Anlage 66
4052 Basel, Switzerland

This is a reprint of articles from the Special Issue published online in the open access journal *Mathematics* (ISSN 2227-7390) (available at: https://www.mdpi.com/journal/metals/special_issues/thermomechanical_processing).

For citation purposes, cite each article independently as indicated on the article page online and as indicated below:

LastName, A.A.; LastName, B.B.; LastName, C.C. Article Title. *Journal Name* **Year**, *Article Number*, Page Range.

ISBN 978-3-03943-380-3 (Hbk)
ISBN 978-3-03943-381-0 (PDF)

© 2020 by the authors. Articles in this book are Open Access and distributed under the Creative Commons Attribution (CC BY) license, which allows users to download, copy and build upon published articles, as long as the author and publisher are properly credited, which ensures maximum dissemination and a wider impact of our publications.

The book as a whole is distributed by MDPI under the terms and conditions of the Creative Commons license CC BY-NC-ND.

Contents

About the Editor vii

Preface to "Applications of Operational Research and Mathematical Models in Management" ix

Petros Kalantonis, Sotiria Schoina, Spyros Missiakoulis and Constantin Zopounidis
The Impact of the Disclosed R & D Expenditure on the Value Relevance of the Accounting Information: Evidence from Greek Listed Firms
Reprinted from: *Mathematics* 2020, 8, 730, doi:10.3390/math8050730 1

Ebenezer Fiifi Emire Atta Mills, Bo Yu and Kailin Zeng
Satisfying Bank Capital Requirements: A Robustness Approach in a Modified Roy Safety-First Framework
Reprinted from: *Mathematics* 2019, 7, 593, doi:10.3390/math7070593 19

Miltiadis Chalikias, Panagiota Lalou and Michalis Skordoulis
Customer Exposure to Sellers, Probabilistic Optimization and Profit Research
Reprinted from: *Mathematics* 2019, 7, 621, doi:10.3390/math7070621 39

Pietro De Giovanni
Digital Supply Chain through Dynamic Inventory and Smart Contracts
Reprinted from: *Mathematics* 2019, 7, 1235, doi:10.3390/math7121235 47

Qian Lei, Juan He and Fuling Huang
Impacts of Online and Offline Channel Structures on Two-Period Supply Chains with Strategic Consumers
Reprinted from: *Mathematics* 2020, 8, 34, doi:10.3390/math8010034 73

Keqing Li, Wenxing Lu, Changyong Liang and Binyou Wang
Intelligence in Tourism Management: A Hybrid FOA-BP Method on Daily Tourism Demand Forecasting with Web Search Data
Reprinted from: *Mathematics* 2019, 7, 531, doi:10.3390/math7060531 93

Vicente Rodríguez Montequín, Joaquín Manuel Villanueva Balsera, Marina Díaz Piloñeta and César Álvarez Pérez
A Bradley-Terry Model-Based Approach to Prioritize the Balance Scorecard Driving Factors: The Case Study of a Financial Software Factory
Reprinted from: *Mathematics* 2020, 8, 276, doi:10.3390/math8020276 107

Pengyu Chen
A Novel Coordinated TOPSIS Based on Coefficient of Variation
Reprinted from: *Mathematics* 2019, 7, 614, doi:10.3390/math7070614 123

Lijun Xu, Yijia Zhou and Bo Yu
Robust Optimization Model with Shared Uncertain Parameters in Multi-Stage Logistics Production and Inventory Process
Reprinted from: *Mathematics* 2020, 8, 211, doi:10.3390/math8020211 141

Ling Liu and Sen Liu
Integrated Production and Distribution Problem of Perishable Products with a Minimum Total Order Weighted Delivery Time
Reprinted from: *Mathematics* 2020, 8, 146, doi:10.3390/math8020146 153

About the Editor

Miltiadis Chalikias is Professor of Statistics and Mathematics at the Department of Accounting and Finance of the University of West Attica. He holds a Ph.D. in Statistics from the National and Kapodistrian University of Athens. He is the author of several scientific articles, books, and chapters. His work is cited by several authors. His principal research interests are in the areas of business mathematics, business statistics, qualitative methods, and multivariate analysis.

Preface to "Applications of Operational Research and Mathematical Models in Management"

During World War I and World War II, many mathematical models were developed and used to solve various optimization problems related to the ongoing war operations. At the end of these wars, these models were applied to industry and business. Nowadays, operational research is a key tool of modern management that is used to solve a wide range of business problems.

The aim of this Special Issue was to discuss new theoretical insights into the applications of operational research and mathematical models in management.

Published papers explore operational research and mathematical models in business management, finance, decision support systems, tourism management, information technology, and artificial intelligence.

Miltiadis Chalikias
Editor

Article

The Impact of the Disclosed R & D Expenditure on the Value Relevance of the Accounting Information: Evidence from Greek Listed Firms

Petros Kalantonis [1,*], Sotiria Schoina [1], Spyros Missiakoulis [2] and Constantin Zopounidis [3,4]

1 Department of Tourism Administration, School of Administrative, Economics and Social Sciences, University of West Attica, 12243 Egaleo, Greece; sot.schoina@yahoo.gr
2 Department of Economics, School of Economics and Political Sciences, National & Kapodistrian University of Athens, 10559 Athens, Greece; s.missiakoulis@gmail.com
3 Financial Engineering Laboratory, School of Production Engineering and Management, Technical University of Crete, 73100 Chania, Greece; kostas@dpem.tuc.gr
4 Finance Department, Audencia Business School, 44312 Nantes, France
* Correspondence: pkalant@uniwa.gr

Received: 31 January 2020; Accepted: 30 April 2020; Published: 6 May 2020

Abstract: Although many empirical studies have focused on R & D performance models for markets globally, the available financial information for R & D expenditure is limited. In other words, can we assume that the reported accounting information for R & D investment is adequate and valuable? This study empirically investigates the effect of R & D reported information on the value relevance of the accounting information of firms' financial statements. Specifically, using Ohlson's equation, it is examined whether changes in stock prices are explained better when R & D factors are included in models, in conjunction with changes in book value and abnormal earnings. We focus on listed firms on the Athens Stock Exchange in order to explore whether R & D expenses are value relevant, in a market which has been affected for a long period by the global economic crisis of 2007. In our findings, we observe that the reported R & D expenses do not have any significant influence on the investors' choices, in contrast to expectations based on the prior literature. Moreover, the panel data analysis employed in the paper overcomes common methodological problems (such as autocorrelation, multicollinearity, and heteroscedasticity) and allows the estimation of unbiased and efficient estimators.

Keywords: value relevance; book value; abnormal earnings; R & D; panel data

1. Introduction

The quotes "all things are flowing", "nothing endures but change", and "nothing stays still" are attributed to Heraclitus of Ephesus (ca. 544–483 BC), who is thought to be the first influential philosopher of change. His theory is that processes of change are important, not the states of rest [1]. However, change and transformation can be accomplished only through innovation. Innovation is inseparably connected to development. Drucker (1985) characterized innovation as a special entrepreneurship tool which contributes to the creation of wealth [2] and Gartner (1990) noted that innovation is one of the factors that constitute the nature of entrepreneurship [3]. Marx (1887) interpreted innovation as the result of companies' attempts to increase their profits [4]. In other words, companies undertake innovation initiatives in the expectation that it will generate a competitive advantage and, thus, significant income from new products and processes.

Subsequently, the term research and development (R & D) is widely linked to innovation. On the one hand, R & D refers to the activities companies undertake to innovate and introduce new products and services. On the other hand, in previous studies, innovation has been recognized as a competitive advantage of firms. As a company gains a competitive advantage by performing in some way that rivals cannot easily replicate, R & D allows that company to remain ahead of its competition. Schumpeter (1934) described how large corporations leave smaller competitors behind through a circular process of positive feedback between innovation and the financing of R & D [5].

The EU, having recognized that innovative firms develop more strategic and organizational skills than non-innovation oriented firms, has formed a specific agenda, underlining the need to invest in research and innovation, by ensuring essential public investment, supporting EU Member States to maximize their R & D expenditure, stimulating private investment, providing a simpler regulatory framework, and supporting innovation procurement [6].

R & D funding is globally examined. The total global expenditure on R & D in 2017 was USD2.2 trillion and continues to grow at a rate of 3.6% per year. The world leader is the U.S., which spends approximately 2.8% of the country's GDP on R & D, while China ranks second, spending almost 1.95% of GDP. Germany is the European leader, spending almost 2.8% of its GDP on R & D. Globally, Greece ranked 51st in 2017, having spent USD 1.83 billion (almost 0.6% of GDP). This is an unexpectedly promising outcome, considering that by 2017 Greece had suffered a severe economic recession for eight consecutive years [7].

According to Hirschey et al. (1985), several reasons explain the differences between a firm's market value and the historical value reported in accounting financial statements [8]. A significant reason is that financial statements are limited to those items that meet the present-day recognition criteria employed by the accounting profession. Thus, potentially relevant items, such as R & D investments, are not reported on balance sheets due to the fact that they do not meet the qualitative criterion of reliability. However, the purpose of financial reports is to provide investors with the information they need to make the best allocation of their investment resources. Indeed, in prior literature, Kalantonis (2011) found evidence that firms' innovative activity affected their performance and, at the same time, was a crucial criterion for investors' decision making [9].

Moreover, Lev et al. (2016) [10] noted that reported financial information has largely lost its relevance. They also proposed that a different accounting treatment of long-term investments in intangibles–such as innovation investments–could improve the value relevance of financial reports. The term "value relevance" reflects the ability of the reported accounting information to explain and summarize the market value of companies (Amir et al., 1993) [11]. Increased relevance means that investors' decisions are based more on the reported accounting information and therefore financial reports become more useful for their users, who can make investment decisions based on reliable and audited information. More effective investment decisions are more necessary during crisis periods.

In this study we explore the effect of R & D disclosed information in firms' financial reports on the value relevance of the reported accounting information. We also investigate the adequateness of the R & D reported financial and non-financial information. Since previous studies have mainly examined the consequences of the lack of reported innovative expenses for the value relevance of accounting information, we contribute to the literature by exploring the effect of the reported financial and non-financial information for the R & D activity of firms, on their annual reports. We focus on both the periods before and during the crisis in order to determine the effect of crisis. This is also a novelty of our study. The examined financial statements in this study are drawn from the listed firms on the Athens Stock Exchange.

It should be noted that the Greek economy has been strongly affected by the financial crisis since 2010. Although a small number of other countries (for example Portugal, Italy and Spain) were also affected by the economic crisis in 2008, the case of Greece is totally different. The crisis in Greece has endured for almost 10 years and can be separated into a number of phases: (i) April 2010—memorandum with International Monetary Fund (IMF), European Central bank (ECB), and European Commission; (ii) June 2015—capital controls; (iii) July 2015—new memorandum with the addition of the European

Stability Mechanism; and (iv) termination of the memorandum and beginning of probation. Moreover, intense political instability prevailed in the country as six legislative elections (October 2009, May 2012, June 2012, January 2015, September 2015, and July 2019) and a referendum (July 2015) took place during the decade of crisis. In no other country did so many events take place that dramatically transformed the entrepreneurial, business, economic, and social environments. In contrast, Cyprus accepted a memorandum in 2013, but in 2015 the first signs of recovery were evident.

Using financial data of the firms listed on the Athens Stock Exchange, we search for significant differences between the value relevance of R & D expenditure compared to the market values. Comparing our findings with those of prior relevant literature, we discuss how the market evaluates the R & D orientation within the framework of the recent Greek financial crisis.

The remainder of the paper is structured as follows. In the following section we explore the research literature and state the research hypothesis. In Section 3 we describe the research methodology. In Section 4 we present the results of the data analysis. The discussion of our findings then follows, and the final section includes the conclusion, the limitations of the research, and future research suggestions.

2. Materials and Methods

2.1. Different Approaches of Measuring R & D Intensity

Examining prior and recent literature, we detected research studies that explored the relationship between innovation or R & D outcomes and the market or financial performance of firms. Nevertheless, other studies investigated the effect of R & D expenditure on firms' value (either book or market value) and their profitability. The findings of these studies were not consistent. As determinant factors of the variability of their results, we recognized the economic status and environment, the type of the studied firms, the conceptual and regulatory approach to the innovation or R & D outcome and investments, and the fact that R & D investments were treated differently in different countries depending on the adopted accounting standards. We categorized these studies according to their approach to R & D measurement and their view of the effects on firms' value or financial performance.

2.1.1. The Non-Monetary Approach

Geroski et al. (1993) evaluated the effects of producing a major innovation on corporate profitability and the differences in profitability between innovators and non-innovators. The study examined the introduction of specific innovations by observing UK manufacturing firms during the period 1972–1983 and showed that the number of innovations produced by a firm has a positive effect on its profitability [12]. Sood et al. (2009) investigated how stock markets react to each event in an innovation project using a sample consisting of U.S. listed firms and collected announcements from 1977 to 2006. Results showed that total market returns to an innovation project were substantially greater than the returns to an average event [13]. Hall et al. (2005) explored the usefulness of patent citations as a measure of the "importance" of a firm's patents [14]. Using patents and citations for 1963–1995, they noted that each citation significantly affected market value, with an extra citation per patent boosting market value by 3%. Szutowski (2016) examined long- and short-term effects of innovation announcements on the market value of the equity of tourism enterprises listed on the 32 most important stock exchanges in the European Union, released during the period of February 2011–February 2016. The study found evidence for positive, statistically significant changes in the market value of the equity of tourism enterprises [15].

2.1.2. The Monetary Approach

In this approach, which is the most commonly used approach by researchers, the R & D intensity is measured as an expenditure, that is, a monetary amount that is either expensed or capitalized. The need for such a distinction has been created because there are two different accounting treatments of R & D expenditure: such an expenditure can either be recorded as an expense in the year it is made, or it can be capitalized and recorded as an asset (under defined conditions and only for development costs, not research costs). By 2005 each country had applied its own rules. Since then, the International Financial Reporting Standards (IFRS) have been implemented, which partly allow capitalization of R & D, but only for development costs and if certain criteria are met. At the same time, many countries apply the Generally Accepted Accounting Principles (GAAP), which differ from the IFRS in their treatment of R & D expenditure (ASC 730) [16]. Under this treatment, R & D costs are recognized as an expense, as they are incurred, since any future economic benefit arising from the development of a given asset is uncertain. In the literature references listed in the remainder of this paper, R & D intensity is measured as an expenditure that is either expensed or capitalized. [17]

A significant amount of recent and relevant literature on the subject of the different accounting approaches of R & D expenditure exists. Gong et al. (2016) investigated whether the nature of differences between national GAAP and IFRS is associated with differential changes in the value relevance of R & D expenses after the adoption of IFRS, using a difference-in-differences study on a sample of public companies in eight European countries and Australia, which covers pre-IFRS and post-IFRS periods during 1997–2012 [18]. They found that the value relevance of R & D expenses declines after IFRS adoption in countries that previously mandated immediate expensing or allowed optional capitalization of R & D costs. On the contrary, they found no change in the value relevance of R & D expenses for countries that switched from the mandatory capitalization rule to IFRS. Chen et al. (2017), having focused on the relevance of voluntary disclosures in a sample of Israeli high-technology and science-based firms, showed that capitalized development costs are highly significant in relation to stock prices [19].

Cazavan-Jeny et al. (2006) tested the value relevance of R & D reporting in a sample of French firms over a 10-year period (1993–2002) and noted that capitalized R & D was significantly negatively associated with stock prices and returns. The authors concluded that this negative coefficient on capitalized R & D implied that investors were concerned with R & D capitalization and reacted negatively to it [20].

2.2. How Does R & D Expenditure Affect the Different Dimensions of Firms' Benchmarks?

2.2.1. R & D Expenditure and Enterprise Performance or Profitability

A large number of surveys are devoted to the way in which R & D intensity impacts an enterprise's performance. Cazavan-Jeny et al. (2011), using a sample of French listed firms for the period 1992–2001, found that firms which capitalized R & D expenditures spend less on R & D and were smaller and poorer performers than those who expensed R & D, showing that the decision to capitalize R & D expenditures is generally associated with a negative impact on future performance. They also showed that when firms both capitalized and expensed R & D expenditures, the expensed portion exhibited a strong negative relationship with future performance [21]. Based on a sample consisting of Australian companies from 1991 to 2001, Chan et al. (2007) suggested that firms with higher R & D intensity perform better, regardless of the accounting method used. Evidence was also found that firms which expense R & D outperform those which capitalize R & D [22]. Cinceraa et al. (2014) examined the sources of Europe's lagging R & D performance relative to the US for the period 2000–2011, and found that young firms in the US succeeded in realizing significantly higher rates of return on R & D compared to their older counterparts, including in high-tech sectors, while European firms failed to generate significant rates of return [23]. Vanderpal et al. (2015) highlighted the nature of the relationship between R & D expense and companies' profitability, having studied firms for a long period, from 1979 to 2013, and obtained evidence supporting a positive relationship between R & D expense and companies' profitability indicators [24]. R & D expense indicators proved to be positively correlated with the profitability of companies (revenues, net income, equity, and Return on Equity). Martin (2015) examined the issue of effectiveness of business innovation and R & D efforts in Polish manufacturing companies, covering the period from 2000 to 2009 [25]. He suggested that positive effects of business R & D are mostly associated with specific time-invariant individual characteristics of business units. Finally, in the most recent study, Turlington et al. (2019), using a sample of firms in the automotive industry in the US and Europe for the period 2006–2016, noted that R & D expenses under U.S. GAAP will be expected to be higher (and income lower) compared to IFRS, as long as absolute R & D costs are growing over time [26]. When growth in R & D investment slows, the R & D expense recorded under U.S. GAAP will begin to approximate the IFRS R & D amounts, since current R & D costs will be closely related to the research expense plus the amortized portion of prior development costs. They also noted that the overall effect on ROE from capitalizing development costs under IFRS is also ambiguous, as the methodology effects both numerator and denominator amounts.

2.2.2. R & D Expenditure and Enterprise's Market Value

A large number of studies have examined how capital markets interpret the information about R & D expenditures disclosed by companies, and these studies mostly find a positive relationship. Lev et al. (1996) addressed the issues of reliability, objectivity, and value-relevance of R & D capitalization by studying US manufacturing companies from 1975 to 1991 and documented the existence of a systematic mispricing of the shares of R & D–intensive companies, or compensation for an extra market risk factor associated with R & D, as they found a significant inter-temporal association between firms' R & D capital and subsequent stock returns [27]. Chambers (2002) searched for differences between the mispricing and risk explanations for R & D-related excess returns in a sample of all NYSE-, ASE-, and NASDAQ-traded firms over the period 1979–1998 and provided convincing evidence of a positive association between the level of R & D investment and post-investment excess stock returns. He proved that the pattern of increasing excess returns to R & D-intensity was associated with risk characteristics of R & D oriented firms [28]. Han et al. (2004) investigated the value-relevance of R & D expenditures of Korean firms from 1988 to 1998, and showed that R & D expenditures were positively associated with stock price [29]. They found a stronger association for the portion of R & D expenditures that was capitalized, rather than expensed. Investors also appeared to interpret fully expensed R & D expenditures as being positive for net present value, however, they suggested that these expenditures should also be capitalized. In a similar study, Ho et al. (2005), using a sample of U.S. firms for an over 40-year period from 1962 to 2001, investigated whether the future share price returns of a firm were positively related to a firm's R & D intensity and showed that R & D investment creates value for firms over one-year and three-year horizons [30]. Ike et al. (2010), using a sample of US firms for the years 1990 through 2007, studied the association between an investment in R & D and market value [31]. They found that the valuation of R & D investment can be linked to a company's market capitalization, in a linear relationship, as investors assess the value relevance of a firm. Başgoze et al. (2013) tested the ability of R & D investment intensity to explain future stock returns, using a sample of enterprises listed on the Istanbul Stock Exchange (ISE) from 2006 to 2010 [32]. Consistent with Lev et al. (1996) and Ike et al. (2010), they showed a linear and statistically significant positive relationship between annual stock returns and R & D investment intensity.

Other studies have contrasting findings to those already mentioned. Chan et al. (1999) investigated whether the stock market appropriately accounts for firms' expenditures on R & D by relating R & D spending to subsequent stock price performance using a sample consisting of all domestic firms listed on the NYSE, AMEX and NASDAQ exchanges from 1975 to 1995. Their evidence did not support a direct link between R & D spending and future stock returns, as the average return over all firms engaged in R & D activity did not differ markedly from that of firms who did not undertake R & D [33]. Callen et al. (2004) found very weak empirical support for the value relevance of R & D expenditures when they investigated a sample selected from the period 1962 to 1996. The study showed that R & D investment significantly affected firm valuation for only 25% of the sample firms [34]. In addition, Sofronas et al. (2019) investigated the relationship between the R & D expenditures and the market value of European companies that reported their annual R & D expenditures consecutively for the years from 2002 to 2012. The study found weak evidence in support of the hypothesis that R & D expenditure positively affects the firm's market value, as well as weak evidence that economic events can disrupt the connection of R & D programs with the market value of firms [35].

In addition, other studies support the view that R & D investments affect profitable and loss-making firms in different ways. Kim et al. (2008) investigated whether there is a non-linear relationship between R & D investments and firm value, using Chinese listed firms between 2005 and 2013 [36]. They showed that R & D investments have an inverted U-shaped relationship with firm value, which indicates that as R & D investments increase, firm value increases to a certain level and then decreases. Franzen et al. (2009) examined whether the valuation relevance of R & D had already been documented for loss-making firms, and extended to profitable firms, by investigating the role of R & D expense in a residual-income based valuation framework across levels of profitability [37]. R & D expense in this study was found to be positively associated with stock prices for loss-making firms and negatively associated with stock prices for profitable firms. In a more recent study, Tsoligkas et al. (2011) examined whether R & D reported assets and expenses were value relevant after the adoption of IFRS in 2005 and searched for any size-related valuation consequences of R & D after IFRS mandatory implementation [38]. They used a sample of UK FTSE listed firms for the years 2006 to 2008. They found evidence to support the view that the capitalized and expensed portions of R & D expenditure are positively and negatively value relevant, respectively, in the UK, after 2005. The hypothesis that there are differences in the valuation of UK companies after the mandatory implementation of IFRS was partially supported with regard to R & D reporting because the expensed portion of R & D was consistently negatively value relevant only for large firms. They finally found that the capitalized portion of R & D was significantly positively related to market value, suggesting that the market perceived these items as successful projects with future economic benefits. In contrast, R & D expenses were significantly negatively related to market values under IFRS, supporting the proposition that they reflected no future economic benefits and thus they should be expensed.

2.3. How Does Market Value Relate to Accounting Data and R & D Information?

Ball et al. (1968) and Beaver (1968) demonstrated the association between abnormal returns and stock prices in the months before and after the dates of earning announcements [39,40]. Then, Hirschey et al. (1985) identified several reasons for the differences between the stock market value and the historical value reported in accounting financial statements [8]. One major reason is that financial statements are limited to those items that meet the present-day recognition criteria employed by the accounting profession. Thus, potentially relevant items such as R & D are not reported on balance sheets because they do not meet the qualitative criterion of reliability. We have already mentioned prior research studies which examined the value relevance of R & D disclosure to the stock market and these studies mostly find a positive relationship between them. Lev et al. (1996) studied the value-relevance of R & D capitalization among US manufacturing companies from 1975 to 1991 found a significantly positive association between firms' R & D capital and subsequent stock returns [27]. Similarly, Han et al. (2004) investigated the value-relevance of R & D expenditures of Korean firms from 1988 to 1998, and found a positive association between R & D expenditures and stock prices [29]. Ike et al. (2010), using a sample of US firms for the years 1990 through 2007, also found a positive, linear relationship between an investment in R & D and market value as investors assessed the value relevance of a firm. Nonetheless, other studies had contrary findings as their evidence did not support a direct link between R & D spending and future stock returns. For example, Callen et al. (2004) found very weak empirical support for the value relevance of R & D expenditures [34]. Finally, studies also exist with mixed findings. For example, Kim et al. (2008) proved an inverted U-shaped relationship between R & D investments and firm value for firms with high growth opportunities, in contrast to firms with low growth opportunities, whose relationship has a plain U-shaped pattern [36]. This corresponds to the study of Franzen et al. (2009), in which R & D expense was shown to be positively associated with stock prices for loss-making firms and negatively associated with stock prices for profitable firms. In addition, the study of Tsoligkas et al. (2011) suggested that the capitalized portion of R & D was significantly positively related to market values, in contrast to the R & D expenses, which were significantly negatively related to market values [38].

2.4. The Crisis Effect

Sofronas et al. (2019), while investigating whether the European economic crisis of 2008 negatively affected the impact of innovation expenditures on the market value of a firm, found weak evidence that such economic events can disrupt the connection of R & D programs with the market value of firms [35]. Ike et al. (2010), among other hypotheses, also investigated how the effect of a global economic disruption, such as 9/11, would negatively affect R & D investment–firm value association. In contrast to Sofronas et al., they proved that disruptive economic events such as 9/11 do impact the scope and effectiveness of R & D investment on firm value.

Hardouvelis et al. (2016) found that the extended period of the economic crisis in Greece has some unique features [41]. Firstly, prior to the crisis little attention was paid to its clear warning signs and pre-existing economic imbalances, despite the fact that such indications had been in place since at least 2006, because the pre-crisis environment was one of rising living standards. Then, crisis consequences developed suddenly in October 2009, when the country's on-going fiscal deficit was discovered to be three times greater than the forecast made a few months earlier, shocking the Eurogroup, rating agencies, and, clearly, markets. Next, the size of the fiscal multiplier was underestimated and labor market reforms were given priority over product market reforms. This had the consequence of worsening the recession, as product prices did not adjust downward immediately, and the drop in nominal wages was translated into a bigger drop in real incomes and domestic aggregate demand. Thereafter, the domestic Greek banks, which had not been affected by the earlier international crisis, saw their capital base completely wiped out when a debt haircut eventually took place in February 2012 and outstanding government bonds and loans were swapped for new bonds. At the end of 2014, when the economy was picking up momentum, the new government who came to power focused on a possible nominal debt haircut. In 2015, three elections were called and a faction supporting Grexit was formed. The population gradually withdrew about EUR 45 bn from banks, accounting for 25% of deposits. Economic sentiment fell drastically, the flow of new investments stopped, and the economy froze. Finally, capital controls were put in place in late June 2015 to prevent further deposit drainage, thus dealing another blow to the private sector and exports. As a result, a third recapitalization of banks took place.

2.5. Hypothesis Statement and Methodology Approach

The main purpose of this research study is to explore the effect of R & D expenditure disclosure on the value relevance of the reported accounting information in the financial statements. Previous and more recent studies, such as these of Franzen et al. (2009), Han et al. (2004), Ike et al. (2010), and Başgoze et al. (2013) [29,31,32,37], focused on the relationship between the R & D expenses and market value of firms. Furthermore, similar studies of Lev et al. (1996) and Tsoligkas et al. (2011) [27,38] investigated the value relevance of R & D investments' capitalization. In addition to these studies, Sofronas et al. (2019) [35] investigated if the relationship between R & D programs and firm value could be affected by an economic crisis.

It is commonly accepted that R & D activity is a key factor for innovation development. Investors are interested in innovative investments, expecting more future benefits from them. Nevertheless, it is important for investors to base their investment choices on reliable information. For that purpose, audited accounting information could be the most appropriate reported information for investors, under the assumption that the disclosed information for the firms' R & D expenditure—in their financial reports—is adequate for investors' decision making. If we accept this assumption, we would expect a significant positive effect of R & D expenditures on the value relevance of financial statements. In other words, the lack of a significant change in value relevance by including R & D in the value relevance equation could be a red flag for the adequacy of accounting information for investors who react positively to firms' R & D orientation.

Many of the previous studies have been influenced by Ohlson's model for the value relevance of accounting information. Ohlson (1995) noted that the value of a firm is equal to the sum of the book value of its equity and the present value of its expected abnormal earnings [42]. Thus, Ohlson's (1995) value relevance model related the stock price to the book value of common equity per share, abnormal earnings per share, and other information. However, Ohlson's model also admits additional information beyond the above accounting metrics, as some value-relevant factors may affect future expected earnings as opposed to current earnings; in other words, accounting measurements incorporate some value-relevant events only after a time delay.

Adopting Ohlson's value relevance equation, in this study we insert additional variables for R & D expenses, R & D disclosure, and economic crises, and we test the following research hypotheses:

Hypothesis 1. *Book value and abnormal earnings are value relevant to market value.*

Hypothesis 2. *A crisis is a determinant factor for the value relevance of the disclosed accounting information.*

Hypothesis 3. *R & D intensity improves the value relevance of the reported accounting information.*

Hypothesis 4. *R & D disclosure effects on the value relevance of the reported accounting information.*

Hypothesis 5. *A crisis affects the value relevance of the accounting information of the firms which disclose information for their R & D activity in their financial reports.*

In this study, we test the significance of the above inserted variables for R & D and crises, and their effect on the value relevance of financial reported information, in order to capture the effect of this additional information, beyond the book value of equity and the present value of expected abnormal earnings.

3. Methodology

3.1. Aims and Scope

The main scope of this paper is to explore the effect of the disclosure of R & D expenses on the value relevance of financial reports. For this purpose, we studied the relevant literature and classified it according to the approach of R & D measurement and the type of the effect of R & D expenditure on firms' value. Next, we selected our sample. All listed firms of the Athens Stock Exchange were included in our sample, excluding financial institutions, banks, and investment and insurance firms, due to the fact that their financial reports have different structures and therefore they are not comparable. Based on the previous literature we stated our hypotheses and defined our variables (dependent, explanatory, and dummy). In this study, we adopted Olson's model, which has been validated in previous studies for valuing firm equity. To avoid problems of endogeneity and autocorrelation in the error terms, we applied panel data regression. Moreover, panel data regression is an appropriate approach for the estimation of microdynamic and macrodynamic effects. Analysis and discussion of data follow, before conclusions, limitations, and further research proposals are stated in the last section of this research.

3.2. Model

The sample consists of all the listed firms on the Athens Stock Exchange. We examined the disclosed accounting information of those firms since 2005, when they adopted IFRS. According to the IFRS framework, R & D activity is discriminated in two phases [43]. Research is the first phase. However, any intangible asset coming from the research activity can be recognized as an expense. Development is the second phase. An intangible asset arising during the phase of development can be recognized only if it meets specific requirements, such as future benefit generation and availability for sale or use. In this paper we focus on the reported R & D expenses. Specifically, R & D expenses divided by Total Assets constitute the R & D intensity, which is specified as one of the added variables in Ohlson's equation.

Ohlson determined the relationship of a firm's market value with accounting variables under three assumptions:

i. The market value of the firm is equal to the present value of all expected future dividends (PVED), assuming non-stochastic interest rates.
ii. A clean surplus relationship is imposed to define the present year book value, which equals the previous year book value plus earnings minus dividends.
iii. Linear information dynamics (which explains the time series behavior of abnormal earnings), establishes a linkage between a firm's intrinsic value and current information [42].

More specifically, the algebraic model can be represented as follows:

$$MV_t = a0 + a_1 \times Bt + a_2 \times AE_t + b_3 \times OI_t + e_i$$

where:

MVt: market value, Bt: book value, AEt: abnormal earnings, OIt: additional information.
In order to test our hypotheses, we formed the following equations:

$$MVS_{it+1} = b_0 + b_1 \times BVS_{it} + b_2 \times AES_{it} + e_{it} \quad (1)$$

$$MVS_{it+1} = b_0 + b_1 \times BVS_{it} + b_2 \times AES_{it} + b_5 \times CRISIS_{it} + e_{it} \quad (2)$$

$$MVS_{it+1} = b_0 + b_1 \times BVS_{it} + b_2 \times AES_{it} + d_t + e_{it} \quad (3)$$

$$MVS_{it+1} = b_0 + b_1 \times BVS_{it} + b_2 \times AES_{it} + b_3 \times RDE_{it} + e_{it} \quad (4)$$

$$MVS_{it+1} = b_0 + b_1 \times BVS_{it} + b_2 \times AES_{it} + b_3 \times RDE_{it} + b_5 \times CRISIS_{it} + e_{it} \quad (5)$$

$$MVS_{it+1} = b_0 + b_1 \times BVS_{it} + b_2 \times AES_{it} + b_3 \times RDE_{it} + d_t + e_{it} \quad (6)$$

$$MVS_{it+1} = b_0 + b_1 \times BVS_{it} + b_2 \times AES_{it} + b_4 \times DISCLOSE_{it} + e_{it} \quad (7)$$

$$MVS_{it+1} = b_0 + b_1 \times BVS_{it} + b_2 \times AES_{it} + b_4 \times DISCLOSE_{it} + b_5 \times CRISIS_{it} + e_{it} \quad (8)$$

$$MVS_{it+1} = b_0 + b_1 \times BVS_{it} + b_2 \times AES_{it} + b_4 \times DISCLOSE_{it} + d_t + e_{it} \quad (9)$$

Specifically, the variables of the equations can be stated as follows:

	Dependent Variable
MVS_{it+1}	Market Value per Share defined as the share price the first day of next year's (t + 1) April.
	Independent Variables
BVS_{it}	Book Value per Share measured as the total common stockholders' equity less the preferred stock, divided by the number of common shares of the company at 31/12 each year.
AES_{it}	Abnormal Earnings per Share at 31/12 defined as the actual earnings per share of current year results (ES_{it}) minus the normal earnings, where normal earnings can be defined as the multiplication of previous year-end book value per share (BVS_{it-1}) and the cost of capital of the firm. As the cost of capital we choose to apply Damodaran's country risk (CR) $AES_{it}: ES_{it} - (CR \times BVS_{it-1})$
RDE_{it}	R & D expenses divided by total assets 31/12
$DISCLOSE_{it}$	Dummy variable indicating whether a firm discloses its R & D expenses or not
$CRISIS_{it}$	Dummy variable of time dividing the study period into two subperiods 2006–2009 and 2010–2017
eit	error

3.3. Methodological Approach

Examining the quantitative methods of the previous studies of value relevance, we observe that researchers have already used time series data and cross-sectional data, and that they have also implemented pooled time-series and cross-sectional regressions [44].

As we have the same 139 cross-sectional units surveyed over a 12-year period, we have balanced panel data. Thus, according to the literature, we adopted panel data analysis by applying a linear regression model. Following this procedure, we avoid the methodological problems of the time-series analysis and the cross-section methods, which often fail to detect the dynamic factors that may affect the dependent variable. In addition, panel data analysis has a number of advantages because it not only provides efficient and unbiased estimators, but also provides a larger number of degrees of freedom available for the estimation, and allows the researcher to overcome the restrictive assumptions of the linear regression model [45].

According to Baltagi (2005) [46], the main difference between time series or cross section regression models and a panel regression model is that the panel regression model has a double subscript on its variables. In the case of our data this would be interpreted as follows:

$$Y_{it} = a + bX_{it} + u_{it}, \; i = 1, \ldots, N \text{ and } t = 1, \ldots, T \tag{10}$$

where i represents the 139 firms of our sample and t represents the 12 years within the period 2006–2017. Consequently, i reflects the cross-section dimension and t reflects the time series dimension of the model. In addition, X_{it} represents the it observation of the five explanatory variables of our equation. In other words, we included the variables X1, X2, and X3 in our model. Thus, our panel data equation becomes:

$$Y_{it} = a + b_1 \times X1_{it} + b2 \times X2_{it} + b3 \times X3_{it} + u_{it} \tag{11}$$

Adopting Baltagi's point of view for error terms in the panel data, we assume that there is a one-way error term, which is uncorrelated with the explanatory variables (Hsiao (2003) [47]) and can be expressed according to Baltagi as follows:

$$U_{it} = \mu_{it} + \nu_{it} \tag{12}$$

The unobservable individual specific effect, which is reflected in μ_{it} and ν_{it}, interprets the usual disturbance in the regression equation. In our specific model—which is based on Ohlson's equation for the measurement of the accounting information's value relevance—the unobservable explanatory variables for the market value can be reflected in u$_{it}$. Hsiao (2003) stated that microdynamic and macrodynamic effects can be estimated with panel data regression analysis and this is its significant advantage [47].

An alternative approach to present the panel regression model was adopted by Karathanasis et al. (2003), who showed that Ohlson's model was superior for the equity valuation [45]. In their study they presented the following approach:

$$Y_{it} = a + \mu_{it} + \lambda_{it} + \sum_{\kappa=1}^{3} (b_\kappa \times X_{\kappa it}) + \varepsilon_{it} \qquad (13)$$

In this model Y_{it} denotes the dependent value for cross section i at time t and $X_{\kappa it}$ denotes the independent (explanatory) variables, which in our research are the book value per share (BVS$_{it}$), the abnormal earnings per share (AES$_{it}$), and the R & D expenses divided by total assets (RDEit). Our dependent variable is market value per share (MVS$_{it+1}$). In this model, μ_i expresses the unobserved cross section effect, λ_t the unobserved time effect, and ε_{it} the remaining non-observed error. We must note that in order to apply Equation (13) we have to assume that either μ_i and λ_i are both fixed or that they are random.

Based on theory and prior literature, we expect a positive and statistically significant effect of both of Ohlson's independent variables (Book Value and Abnormal Earnings on the Market Value). However contradictory findings have been detected in previous studies regarding the effect of the R & D and crises on firms' value.

3.4. Data and Descriptives

Our sample consists of 139 firms listed on the Athens Stock Exchange for the period 2006–2017, as the IFRS were adopted in Greece in 2005. Accounting data were collected from firms' annual balance sheets and financial statements. Stock prices were retrieved from internet [48]. To be included in the sample, necessary accounting and market data must have been available. In addition, the sample was confined to firms with December fiscal year-ends. Banks, financial, assurance, and real estate companies was excluded. The exact sample size was 1668 total observations. The market value of common stock (MV$_{it+1}$) was as of the first day of April in year t + 1. This allows a 3-month filing period for year t financial statements, to ensure that market value is measured after the release of the information.

In order to explore the impact of financial crisis on the value relevance of accounting information for firms which disclose their R & D expenses, we divided the financial data into two periods. The cut-off year for the division of the two subperiods is 2010. In 2010, the Greek economy began financial probation of the EU and the IMF. Therefore, we considered the period from 2010 to 2017 as the crisis period and the period from 2006 to 2009 as the pre-crisis period for the Greek economy.

Descriptive statistics of the equation's variables are presented in Table 1. We observe that the Greek listed firms spend annually, on average, 0.2% of their total asset value for their R & D activities. Since the mean, minimum, and maximum values of R & D intensity were not significantly charged after the beginning of the Greek economy probation period, we could consider that the crisis did not affect the R & D intensity of the Greek listed firms. However, 28.24% of the firms disclosed information for their R & D expenses in their annual financial reports.

Table 1. Descriptive statistics.

	Mean	Median	St. Deviation	Min.	Max
MVS_{t+1}	4.1079	1.0450	15.6100	0.0060	286.0000
BVS_{it}	3.3504	1.5405	11.3500	−15.7550	261.6800
AES_{it}	−0.0193	−0.1012	4.6583	−12.9500	138.7900
RDE_{it} total	0.0026	0.0000	0.0087	0.0000	0.0858
RDE_{it} before	0.0028	0.0000	0.0099	0.0000	0.0858
RDE_{it} during	0.0025	0.0000	0.0081	0.0000	0.07846

4. Results and Discussion

This study attempts to explore the impact of R & D reported expenses on the value relevance of accounting information. Previous similar studies used Ohlson's equation to measure the value relevance of financial statements. The main two components introduced by Ohlson to measure the relevance are the book value and the abnormal earnings. Both can be noted as independent variables of the regression equation. Indeed, the market value is the dependent variable of Ohlson's model for the measurement of value relevance.

The quantitative approach proposed in prior and recent literature for regression analysis is an OLS regression for pooled data. We applied the F-test for fixed effects, from which we assumed that the fixed effect model is better than the pooled OLS. We also applied the Breusch–Pagan LM test for random effects, from which we assumed that the random effect model is able to deal with heterogeneity better than the pooled OLS. Then we applied the Hausman test for comparing fixed and random effects. From this test we assumed that the random effect model is able to deal with heterogeneity better than the pooled OLS. However, due to the fact that we have time-series and cross-sectional data, we also tested the results of the regression analysis for heteroscedasticity. We selected the White test, which is an appropriate test for heteroscedasticity [49]. In the results shown in Table 2 we observe that there is no significant evidence to accept the null hypothesis, which has been stated as follows: there is no heteroscedasticity when all the coefficients are equal to zero. The chi-square value obtained (X^2 = 1618) exceeds the critical chi-square value $p(X^2)$ at the chosen level of significance and therefore the p-value is approximately zero. Then, according to the findings, we cannot assume homoscedasticity and we implement WLS panel data analysis.

Table 2. White's OLS heteroscedasticity test.

Independent Variables	Coefficient	St. Deviation	t-Statistics	p-Value
b0	56.9886	9.3988	6.06	<0.0001 ***
BVS_{it}	−28.5945	1.8487	−15.47	<0.0001 ***
AES_{it}	−30.24	8.8765	−3.4	0.0007 ***
$sqBVS_{it}$	1.0495	0.0118	88.45	0.0000 ***
$X2X3_{it}$	−0.8522	0.1228	−6.94	<0.0001 ***
$sqAES_{it}$	0.4393	0.0635	6.91	<0.0001 ***
R-square Adjusted		0.9703		
chi-square		1618.549		

*** 1%, ** 5%, * 10% significance level.

As shown in Table 3, the book value and the abnormal earnings of the examined firms have a significant effect on their market value. The R^2 is approximately 0.40, which indicates an adequate level of relevance. Of course, this implies unexplained variability of almost 60%. As we have already mentioned, the examined period was divided into two sub-periods. The first period was before the beginning of the Greek economy's financial probation, and the second was the period during the probation. Introducing to Ohlson's equation a variable for the crisis, we observe a 5% increase in adjusted R^2. Looking at Model 2 of Table 3, we can observe, first, that the book value and the abnormal earnings positively affect the market value, and, second, that the crisis significantly negatively affects the market value.

Table 3. Weighted least squares regression analysis.

Independent Variables	Coefficient	St. Deviation	t-Statistics	p-Value
Model 1				
b_0	0.4943	0.0417	11.84	<0.0001 ***
BVS_{it}	0.6535	0.0198	32.9	<0.0001 ***
AES_{it}	0.2059	0.0519	3.96	<0.0001 ***
R-square Adjusted			0.4070	
Model 2				
b_0	1.3371	0.0712	18.78	<0.0001 ***
BVS_{it}	0.6282	0.0203	30.84	<0.0001 ***
AES_{it}	0.1092	0.0501	2.17	0.0295 **
$CRISIS_{it}$	−1.1207	0.0774	−14.47	<0.0001 ***
R-square Adjusted			0.4507	
Model 3				
b_0	0.6114	0.133	4.45	<0.0001***
BVS_{it}	0.6118	0.02	30.52	<0.0001 ***
AES_{it}	0.0794	0.0481	1.65	0.0989 *
$d2006_t$	1.937	0.1852	10.46	<0.0001 ***
$d2007_t$	1.4109	0.1852	7.61	<0.0001 ***
$d2008_t$	−0.0814	0.1848	−0.44	0.6595
$d2009_t$	0.0306	0.1846	0.16	0.8681
$d2010_t$	−0.4124	0.1846	−2.23	0.0257 **
$d2011_t$	−0.7089	0.1849	−3.83	0.001 ***
$d2012_t$	−0.4093	0.1849	−2.21	0.0270 **
$d2013_t$	−0.0406	0.185	−0.21	0.8262
$d2014_t$	−0.4103	0.1846	−2.22	0.0264 **
$d2015_t$	−0.4596	0.1845	−2.49	0.0129 **
$d2016_t$	−0.3971	0.1845	−2.15	0.0315 **
R-square Adjusted			0.4825	
Model 4				
b_0	0.4865	0.0431	11.29	<0.0001 ***
BVS_{it}	0.6526	0.0198	32.81	<0.0001 ***
AES_{it}	0.2041	0.0519	3.92	<0.0001 ***
RDE_{it}	3.9956	4.5097	0.88	0.3757
R-square Adjusted			0.4056	
Model 5				
b_0	1.3302	0.0725	18.33	<0.0001 ***
BVS_{it}	0.6278	0.0204	30.76	<0.0001 ***
AES_{it}	0.1072	0.0501	2.13	0.0326 **
RDE_{it}	2.1497	4.3043	0.49	0.6175
$CRISIS_{it}$	−1.1203	0.0775	−14.45	<0.0001 ***
R-square Adjusted			0.449	
Model 6				
b0	0.6059	0.1345	4.5	<0.0001 ***
BVS_{it}	0.6125	0.0201	30.35	<0.0001 ***
AES_{it}	0.078	0.0481	1.62	0.1052
RDE_{it}	1.3561	3.9447	0.34	0.731
$d2006_t$	1.9269	0.1856	10.38	<0.0001 ***
$d2007_t$	1.411	0.1855	7.6	<0.0001 ***

Table 3. Cont.

Independent Variables	Coefficient	St. Deviation	t-Statistics	p-Value
d2008$_t$	−0.0823	0.1852	−0.44	0.6568
d2009$_t$	0.0291	0.185	0.16	0.8747
d2010$_t$	−0.4125	0.1851	−2.23	0.260 **
d2011$_t$	−0.7113	0.1853	−3.84	0.001 ***
d2012$_t$	−0.4097	0.1853	−2.21	0.0272 **
d2013$_t$	−0.04	0.1853	−0.21	0.8291
d2014$_t$	−0.4154	0.185	−2.24	0.0249 **
d2015$_t$	−0.4632	0.1849	−2.5	0.0124 **
d2016$_t$	−0.4013	0.1849	−2.17	0.0301 **
R-square Adjusted		0.4806		
Model 7				
b$_0$	0.4492	0.0452	9.92	<0.0001 ***
BVS$_{it}$	0.6442	0.0201	32	<0.0001 ***
AES$_{it}$	0.2055	0.0518	3.96	<0.0001 ***
DISCLOSE$_{it}$	0.221	0.0769	2.87	0.0041 ***
R-square Adjusted		0.4087		
Model 8				
b$_0$	1.301	0.073	17.82	<0.0001 ***
BVS$_{it}$	0.6227	0.0204	30.47	<0.0001 ***
AES$_{it}$	0.1082	0.0501	2.16	0.0308 **
DISCLOSE$_{it}$	0.1888	0.0744	2.53	0.0112 **
CRISIS$_{it}$	−1.1487	0.0776	−14.8	<0.0001 ***
R-square Adjusted		0.4537		
Model 9				
b$_0$	0.5328	0.1359	3.91	<0.0001 ***
BVS$_{it}$	0.6106	0.02	30.5	<0.0001 ***
AES$_{it}$	0.0786	0.048	1.63	0.1022
DISCLOSE$_{it}$	0.2252	0.0788	2.85	0.0043
d2006$_t$	1.9415	0.1852	10.48	<0.0001 ***
d2007$_t$	1.4201	0.1851	7.66	<0.0001 ***
d2008$_t$	−0.0699	0.1848	−0.37	0.7051
d2009$_t$	0.0397	0.1846	0.21	0.8294
d2010$_t$	−0.3981	0.1846	−2.15	0.0312 **
d2011$_t$	−0.705	0.1848	−3.81	0.001 ***
d2012$_t$	−0.4104	0.1849	−2.22	0.0266 **
d2013$_t$	−0.0425	0.1849	−0.22	0.8182
d2014$_t$	−0.4173	0.1846	−2.26	0.0239 **
d2015$_t$	−0.4641	0.1845	−2.51	0.0120 **
d2016$_t$	−0.4006	0.1844	−2.17	0.0300 **
R-square Adjusted		0.4867		

*** 1%, ** 5%, * 10% significance level.

The placement of the Greek economy under the status of financial probation occurred in May of 2010. According to our findings (Model 3, Table 3), all the years of the period 2010–2016 significantly negatively affected the market value of firms as also shown by other studies [50,51]. Different findings have been observed regarding the effect of R & D expenses on firms' market value, both prior to and under probation. Specifically, there is no evidence of a significant impact of R & D expenses on firms' market value. Furthermore, we should note that the insertion of the variable R & D expenses in Ohlson's equation had no effect on the adjusted R^2 (Model4, Table 3). Based on the above findings we cannot claim that R & D expenses are relevant to market value. On the other hand, as is shown (Model 7, Table 3), the disclosure of R & D affirms activity and significantly positively affects the market value. Nevertheless, it does not seem to improve the fit of the model, since no significant change in adjusted R^2 is observed.

5. Concluding Remarks

This study explored the impact of the reported R & D expenditure on the value relevance of financial statements. We measured the R & D expenditure under the assumption that it is reflected in the R & D disclosed expenses. Ohlson's model was adopted for the estimation of the value relevance of the disclosed accounting information. We imported, in addition to the typical variables of Ohlson's equation, R & D intensity and R & D disclosure as independent variables in the model. The research period was divided into the "pre-crisis" and "during the crisis" periods. Two different quantitative analysis approaches were implemented in order to test our research hypotheses. Although OLS regression analysis has been used in previous studies, we applied the WLS panel data regression analysis to avoid heteroscedasticity. The implementation of this method and the fact that the research target was Greek listed firms for the period 2006–2017, which have not been analyzed in previous relevant studies, constitutes our contribution to the research literature. We must stress that the Greek economy was affected by the global crisis later than European countries. Nevertheless, the duration of the consequences of the global crisis was extremely long compared to those of other European countries.

In this research paper we documented evidence of the effects of R & D disclosed information on the value relevance of the reported financial information. We positively verified Hypothesis 1 and Hypothesis 4, and negatively verified Hypothesis 2 and Hypothesis 5, but found no verification for Hypothesis 3. Next, the basic model of Ohlson was also positively verified in our study. In addition, our results showed that R & D expenses were not value relevant to the market value, but the disclosure of R & D was positively value relevant to the market value. An interpretation of these findings is that investors are interested in firms which report their R & D activity, but the reported amount does not seem to be of interest. Another finding to be highlighted is that the financial crisis significantly negatively affected the market value of Greek listed firms. However, we did not find evidence to prove that the financial crisis was a determinant factor of value relevance. An interpretation of our findings could be that the reported R & D expenditure is not sufficiently adequate to allow investors to make investment decisions. We believe that the managers of firms are not willing to disclose more financial information than is required according to the legal and regulatory framework. In this study, we highlight the necessity of an improvement of the legal framework in the direction of an obligatory reporting of capitalized R & D information, which could be more attractive to the investors and shareholders.

The results of our investigation are in accordance with Zhao's (2002) study [52], as we both found evidence to support the view that R & D reporting has a significant effect on the association of equity price with accounting data. Concerning the fact that we found no evidence to support the view that R & D expenses are value relevant to market value, we verify previous studies, such these of Chan et al. (1999), Callen et al. (2004), and Sofronas et al. (2019) [22,34,35]. Finally, our findings are in agreement with the literature showing that investors react positively to capitalized R & D investment, while they are indifferent to, or negatively placed against, expended R & D.

It is clear that R & D input affects various factors, for example, strategic alliances and external investments, until it enhances financial performance. This is because R & D input represents a firm's willingness to invest in technology. At the same time, R & D may affect the market reputation of a firm [53–55]. Therefore, it is inevitable to create various direct and indirect paths from R & D input to financial outcomes. Nevertheless, it is not possible to measure this supplementary financial outcome, as Greek firms release little relevant information. Thus, this is a limitation of our study. We must also note that we used data of the Greek Stock Exchange and this could be another limitation in our study. Moreover, the fact that we did not expand our research to include the period before 2006 could also be a limitation of our study. This is because the financial data before 2005 were reported according to the Greek accounting standards, and not the international standards that were adopted in 2005.

Future research could be extended to all the PIGS countries. Furthermore, the fact that R & D disclosed expenditures are not relevant to market value could support further discussion with the IASB or the national boards for accounting standards.

Author Contributions: P.K. contributed to the paper's conceptualization. S.S. developed and designed the research methodology and models. S.M. performed the experiments, calculations and computer programming. C.Z. supervised. All authors have read and agreed to the published version of the manuscript.

Funding: This research received no external funding.

Conflicts of Interest: The authors declare no conflicts of interest.

References

1. Müller-Merbach, H. Heraclitus: Philosophy of change, a challenge for knowledge management? *Knowl. Manag. Res. Pract.* **2006**, *4*, 170–171. [CrossRef]
2. Drucker, P. *Innovation and Entrepreneurship*; Harper & Row: New York, NY, USA, 1985.
3. Gartner, W. What are we talking about when we talk about entrepreneurship? *J. Bus. Ventur.* **1990**, *5*, 15–28. [CrossRef]
4. Kautsky, K.; Stenning, H.J. *The Economic Doctrines of Karl Marx*; N.C.L.C. Publishing Society: London, UK, 1925.
5. Schumpeter, J. *Theory of Economic Development*; Harvard Economics Studies: Cambridge, MA, USA, 1934; Volume 46.
6. Eurostat. *Regional Yearbook*; Publications Office of the European Union: Luxembourg, 2019; p. 120.
7. Available online: https://www.rdworldonline.com (accessed on 20 January 2020).
8. Hirschey, M.; Weygandt, J. Amortization Policy for Advertising and Research and Development Expenditures. *J. Account. Res.* **1985**, *23*, 326–335. [CrossRef]
9. Kalantonis, P. Financial Statements and Recent Needs for Financial Information. The Case of Innovation Investments. Ph.D. Thesis, University of West Attica, Chania, Greece, 2011.
10. Lev, B.; Gu, F. *The End of Accounting and the Path Forward for Investors and Managers*; John Wiley & Sons, Inc.: Hoboken, NJ, USA, 2016.
11. Amir, E.; Harris, T.S.; Venuti, E.K. A comparison of the value-relevance of US versus non-US GAAP accounting measures using form 20-F reconciliations. *J. Account. Res.* **1993**, *31*, 230–264. [CrossRef]
12. Geroski, P.; Machin, S.; Van Reenen, J. The Profitability of Innovative Firms. *RAND J. Econ.* **1993**, *24*, 198–211. [CrossRef]
13. Sood, A.; Tellis, G. Do Innovations Really Pay Off? Total Stock Market Returns to Innovation. *Mark. Sci.* **2009**, *28*, 442–456. [CrossRef]
14. Hall, B.; Jaffe, A.; Trajtenberg, M. Market Value and Patent Citations. *RAND J. Econ.* **2005**, *36*, 16–38.
15. Szutowski, D. *Innovation & Market Value, the Case of Tourism Enterprises*; Difin SA: Warszawa, Poland, 2016.
16. Available online: https://www.irs.gov/pub/irs-utl/gbc_p_272_07_01_02.pdf (accessed on 10 January 2020).
17. Available online: http://eifrs.ifrs.org/eifrs/bnstandards/en/IAS38.pdf (accessed on 10 January 2020).
18. Gong, J.J.; Wang, S.L. Changes in the value relevance of research and development expenses after IFRS adoption. *Adv. Account.* **2016**, *35*, 49–61. [CrossRef]
19. Chen, E.; Gavious, I.; Lev, B. The positive externalities of IFRS R&D capitalization: Enhanced voluntary disclosure. *Rev. Acc. Stud.* **2017**, *22*, 677–714.
20. Cazavan-Jeny, A.; Jean, T. The negative impact of R&D capitalization: A value relevance approach. *Eur. Account. Rev.* **2006**, *15*, 37–61.
21. Cazavan-Jeny, A.; Jeanjean, T.; Joos, P. Accounting choice and future performance: The case of R&D accounting in France. *J. Account. Public Policy* **2011**, *30*, 145–165.
22. Chan, H.; Faff, R.; Gharghori, P.; Ho, Y.K. The relation between R&D intensity and future market returns: Does expensing versus capitalization matter? *Rev. Quant. Financ. Account.* **2007**, *29*, 25–51.
23. Cincera, M.; Veugelers, R. Differences in the rates of return to R&D for European and US young leading R&D firms. *Res. Policy* **2014**, *43*, 1413–1421.
24. VanderPal, G. Impact of R&D Expenses and Corporate Financial Performance. *J. Account. Financ.* **2015**, *15*, 135–149.

25. Martin, M. Effectiveness of Business Innovation and R&D in Emerging Economies: The Evidence from Panel Data Analysis. *J. Econ. Bus. Manag.* **2015**, *3*, 440–446.
26. Turlington, J.; Fafatas, S.; Goad Oliver, E. Is it U.S. GAAP or IFRS? Understanding how R&D costs affect ratio analysis. *Bus. Horizons* **2019**, *62*, 427–436.
27. Lev, B.; Sougiannis, T. The capitalization, amortization, and value-relevance of R&D. *J. Acount. Econ.* **1996**, *21*, 107–138.
28. Chambers, D.; Jennings, R.; Thompson, R. Excess Returns to R&D-Intensive Firms. *Rev. Account. Stud.* **2002**, *7*, 133–158.
29. Han, B.H.; Manry, D. The value-relevance of R&D and advertising expenditures: Evidence from Korea. *Int. J. Account.* **2004**, *39*, 155–173.
30. Ho, Y.K.; Keh, H.T.; Ong, J.M. The Effects of R&D and Advertising on Firm Value: An Examination of Manufacturing and Nonmanufacturing Firms. *IEEE Trans. Eng. Manag.* **2005**, *52*, 3–14.
31. Ike, C.E.; Kingsley, O. The effect of R&D investment on firm value: An examination of US Manufacturing and service industries. *Int. J. Prod. Econ.* **2010**, *128*, 127–135.
32. Başgoze, P.; Sayin, H.C. The effect of R&D expenditure (investments) on firm value: Case of Istanbul stock exchange. *J. Bus. Econ. Financ.* **2013**, *2*, 5–12.
33. Chan, L.; Lakonishok, J.; Sougiannis, T. *The Stock Market Valuation of Research and Development Expenditures*; NBER Working Paper No. 7223; National Bureau of Economic Research: Cambridge, MA, USA, 1999.
34. Callen, J.L.; Morel, M. The valuation relevance of R&D expenditures: Time series evidence. *Int. Rev. Financ. Anal.* **2005**, *14*, 304–325.
35. Sofronas, C.; Archontakis, F.; Smart, P. Decision making under uncertainty? R&D activity and market value during financial crisis. *Eur. J. Innov. Manag.* **2019**, *23*, 3.
36. Kim, W.S.; Park, K.; Lee, S.H.; Kim, H. R&D Investments and Firm Value: Evidence from China. *Sustainability* **2018**, *10*, 4133.
37. Franzen, L.; Radhakrishnan, S. The value relevance of R&D across profit and loss firms. *Acc. Public Policy* **2009**, *28*, 16–32.
38. Tsoligkas, F.; Tsalavoutas, I. Value relevance of R&D in the UK after IFRS mandatory implementation. *Appl. Financ. Econ.* **2011**, *21*, 957–967.
39. Ball, R.; Brown, P. An Empirical Evaluation of Accounting Income Numbers. *J. Account. Res.* **1968**, *6*, 159–178. [CrossRef]
40. Beaver, W. The Information Content of Annual Earnings Announcements Reviewed work(s). *J. Account. Res.* **1968**, *6*, 67–92. [CrossRef]
41. Hardouvelis, G.; Gkionis, I. A Decade Long Economic Crisis: Cyprus versus Greece. *Cyprus Econ. Policy Rev.* **2016**, *10*, 3–40.
42. Ohlson, J. Earnings, book values and dividends in equity valuation. *Contemp. Account. Res.* **1995**, *11*, 661–687. [CrossRef]
43. Available online: http://www.icab.org.bd/icabweb/webNewsEventNoticeCir/viewPdf?fileWithPath=/app/share_Storage/Attachments/icabwebcommonupload/images/upload/webupload/general_file/general_file/IAS_38_2017.pdf (accessed on 10 January 2020).
44. Kumari, P.; Mishra, C.S. A Literature Review on Ohlson. *Asian J. Financ. Account.* **1995**. [CrossRef]
45. Karathanassis, G.; Spilioti, S. An Empirical Investigation of the Traditional and the Clean Surplus Valuation Models. *Manag. Financ.* **2003**, *29*, 55–66. [CrossRef]
46. Baltagi, B. *Econometric Analysis of Panel Data*; John Wiley & Sons Ltd.: England, UK, 2005.
47. Available online: www.naftemporiki.gr (accessed on 10 May 2019).
48. Hsiao, C. *Analysis of Panel Data*; Cambridge University Press: Cambridge, UK, 2003.
49. Gujarati, D.N. *Basic Econometrics*, 4th ed.; Tata McGraw-Hill Publishing Company Limited: New Delhi, India, 2004.
50. Balios, D.; Daskalakis, N.; Eriotis, N.; Vasiliou, D. SMEs capital structure determinants during severe economic crisis: The case of Greece. *Cogent Econ. Financ.* **2016**, *4*. [CrossRef]
51. Chalikias, M.; Lalou, P.; Skordoulis, M.; Papadopoulos, P.; Fatouros, S. Bank oligopoly competition analysis using a differential equations model. *Int. J. Op. Res.* **2020**, *38*, 137–145. [CrossRef]
52. Zhao, R. Relative Value Relevance of R&D Reporting: An International Comparison. *J. Int. Financ. Manag. Account.* **2002**, *13*, 153–174.

53. Chalikias, M.; Skordoulis, M. Implementation of FW Lanchester's combat model in a supply chain in duopoly: The case of Coca-Cola and Pepsi in Greece. *Oper. Res.* **2017**, *17*, 737–745.
54. Acs, Z.J.; Audretsch, D.B.; Feldman, M.P. R & D spillovers and recipient firm size. *Rev. Econ. Stat.* **1994**, *76*, 336–340.
55. Artz, K.W.; Norman, P.M.; Hatfield, D.E.; Cardinal, L.B. A longitudinal study of the impact of R&D, patents, and product innovation on firm performance. *J. Prod. Innov. Manag.* **2010**, *27*, 725–740.

© 2020 by the authors. Licensee MDPI, Basel, Switzerland. This article is an open access article distributed under the terms and conditions of the Creative Commons Attribution (CC BY) license (http://creativecommons.org/licenses/by/4.0/).

Article

Satisfying Bank Capital Requirements: A Robustness Approach in a Modified Roy Safety-First Framework

Ebenezer Fiifi Emire Atta Mills [1,2], Bo Yu [3] and Kailin Zeng [1,2,*]

[1] Department of Finance, School of Economics & Management, Jiangxi University of Science & Technology, Ganzhou 341000, China
[2] Ganzhou Academy of Financial Research (GAFR), Ganzhou 341000, China;
[3] School of Mathematical Sciences, Dalian University of Technology, Dalian 116024, China; yubo@dlut.edu.cn
* Correspondence: kailinzeng12@163.com

Received: 20 May 2019; Accepted: 28 June 2019; Published: 1 July 2019

Abstract: This study considers an asset-liability optimization model based on constraint robustness with the chance constraint of capital to risk assets ratio in a safety-first framework under the condition that only moment information is known. This paper aims to extend the proposed single-objective capital to risk assets ratio chance constrained optimization model in the literature by considering the multi-objective constraint robustness approach in a modified safety-first framework. To solve the optimization model, we develop a deterministic convex counterpart of the capital to risk assets ratio robust probability constraint. In a consolidated risk measure of variance and safety-first framework, the proposed distributionally-robust capital to risk asset ratio chance-constrained optimization model guarantees banks will meet the capital requirements of Basel III with a likelihood of 95% irrespective of changes in the future market value of assets. Even under the worst-case scenario, i.e., when loans default, our proposed capital to risk asset ratio chance-constrained optimization model meets the minimum total requirements of Basel III. The practical implications of the findings of this study are that the model, when applied, will provide safety against extreme losses while maximizing returns and minimizing risk, which is prudent in this post-financial crisis regime.

Keywords: robust optimization; capital to risk asset ratio; chance constraint; safety-first principle; Basel III; capital requirements

---

1. Introduction

Undoubtedly, after the Global Financial Crisis (GFC) of 2007–2008, the spotlight on capital requirement heightened. The Global Financial Crisis was caused by "sub-prime" housing loans in the form of mortgage-backed securities. Numerous determinants for the Global Financial Crisis have been proposed, with various weights assigned by researchers [1]. The Financial Crisis Inquiry Commission stated that the Global Financial Crisis was avertable and its root cause was "widespread failures in financial regulation and supervision...". A persistent deficit of bank capital and a 25% dip in private investment on average comprise some of the aftermaths of GFC [2]. In tackling the problems and loopholes in financial regulations unveiled by the GFC, Basel III was proposed. Its main objective is to bolster bank capital requirement by expanding bank liquidity and lessening bank leverage. The changes in Basel III include a meaningful surge in the Capital to Risk (weighted) Assets Ratio (CRAR) [3].

The utilization of CRAR safeguards depositors and improves the efficiency and stability of financial frameworks. In [4], a CRAR chance-constrained optimization model was proposed to guarantee that a bank can cope with the capital requirements of Basel III with a probability of 95%, irrespective of the changes in the future market value of assets. The proposed model considered loans having truncated Gaussian distributed returns, which allowed reformulating the chance constraint related to capital requirements in a second-order cone condition. For the purpose of completeness, the CRAR chance constraint is re-introduced:

$$\mathbb{P}\left\{y^0(x) + y(x)^T \zeta \leq 0\right\} \geq \alpha, \tag{1}$$

where $y^0(x) = TL - DR - Rr - LP - M\sum_{k=u+1}^{u+v}(1+\mathbf{R}_k)x_k$ and $y(x) = M\sum_{k=1}^{u}(\lambda\omega_k - 1)x_k$. TL is the bank's total liability, and M is the bank's total asset amount. For the purpose of this study, Tier 1 capital (core capital) consists of shareholders' equity and disclosed reserves, DR. Tier 2 (supplementary capital) consists of revaluation reserves, Rr, and general loan loss provisions, LP. Denote $\mathbf{R} = [\mathbf{R}_1, \mathbf{R}_2, \ldots, \mathbf{R}_u, \mathbf{R}_{u+1}, \ldots, \mathbf{R}_{u+v}]^T$ as the vector of the annual interest rate of loans and the treasury bill, fixed assets, and non-interest earning assets (riskless). $\Omega = [\zeta^T, \xi^T]^T$ is the vector of assets with $\zeta = [\zeta_1, \zeta_2, \ldots, \zeta_u]^T$ and $\xi = [\xi_1, \xi_2, \ldots, \xi_v]^T$ corresponding to loans and riskless assets (treasury bill, fixed assets, and non-interest earning assets), respectively. ζ constitutes uncertain parameters that can be estimated, and ξ is a deterministic vector of $[1 + \mathbf{R}_{u+1}, 1 + \mathbf{R}_{u+2}, \ldots, 1 + \mathbf{R}_{u+v}]^T$. The Basel III total capital requirement ratio is denoted as λ; ω_k is the kth asset's weight factor; and α is the safety factor. Denote $x = [x_1, x_2, \ldots, x_u, x_{u+1}, \ldots, x_{u+v}]^T$ as the vector of asset allocation or investment proportion, which is the decision variable.

The work in [4] assumed that the full and accurate probability distribution of the random vector ζ is known, given estimations from historical data and information from the literature. However, one might have only partial information about the probability distribution: its moment information. Therefore, replacing an unknown distribution with a particular distribution might lead to an over-optimistic solution, resulting in an unsatisfactory chance constraint under the true or actual distribution of random vector ζ. The work in [5] stated that a more difficult challenge that arises is the need to commit to a particular distribution of random vector ζ given only restricted information about the stochastic parameter. To avoid the difficulty of selecting a proper distribution and uncertainty surrounding it, the work in [6] explored the distributionally-robust optimization approach. In this approach, after defining a set $\mathcal{P}$ of possible probability distributions that are assumed to include the true probability distribution $\mathcal{D}$ of random vector ζ, the optimization problem is reconstructed with respect to the worst case expected function over the selection of the probability distribution in this set. Uncertainty in parameter ζ is described through uncertainty sets that contain many possible values realized for random vector ζ. When the uncertainty set is characterized by statistical estimates of the mean and covariance, the work in [7] provided a sufficient condition to guarantee the satisfaction of the constraint with distribution uncertainty at a specified confidence level.

A natural way to tackle a chance constraint against parameter uncertainty is to use the constraint robustness approach. In particular, the distributionally-robust CRAR chance constraint can be expressed as:

$$\inf_{\mathbb{P}\in\mathcal{P}} \mathbb{P}\{y^0(x) + y(x)^T\zeta \leq 0\} \geq \alpha, \tag{2}$$

where $\mathcal{P}$ denotes the set of all probability distributions that are consistent with the first and second moments of the probability distribution of the random vector, ζ. Whenever x satisfies (2) and $\mathcal{D} \in \mathcal{P}$ is the true distribution, x satisfies the chance constraint (1) under true probability distribution $\mathcal{D}$. The work in [8] revealed that contrary to the stochastic programming approach, the distributionally robust chance constraint reflects investors' risk and aversion towards exposure to uncertainty about the probability distribution of the outcomes via consideration of the worst probability distribution within $\mathcal{P}$. Thus, this study aims to close the research gap by employing the distributionally-robust approach as a way to avoid the difficulty of selecting a proper distribution and uncertainty surrounding the random vector in the framework of meeting capital requirements.

Empirical evidence indicates that the failures of several banks during the GFC kindled concern about maintaining the excessive risk-taking behaviour of banks. Thus, employing variance-Roy's safety-first risk measure as a way of minimizing risk while providing a safety net against extreme losses is reasonable and worth investigating. Therefore, we employed the modified and improved

Roy's safety-first principle investigated by [9]. It is important to note that appropriately modelling risk and meeting capital requirements among other objectives are important for financial stability and that an economy with an efficient financial market structure develops faster.

This paper aims to extend the proposed single-objective CRAR chance constrained optimization model in [4] by considering the multi-objective constraint robustness approach in a modified safety-first framework. This paper also considers credit risk and the expected value of the portfolio. In solving the model, Section 3 introduces steps in constructing a deterministic convex counterpart of robust the probability constraint (2).

In summary, this paper considers a multi-objective distributionally-robust chance-constrained model for capital adequacy. A deterministic equivalent of the robust chance constraint is developed, and computational results are provided to suggest that the model is effective at generating capital allocation decisions even under the worst case realizations (i.e., default) of the debt instruments considered.

The structure of this study is organized as follows: in the next sections, literature review on optimization under uncertainty is discussed, and problem definition and assumption are presented. Section 4 provides the model formulation and approach, and the next section presents the development of the model. Numerical examples and computational results of our method are shown in Section 6. The last section concludes the paper.

2. Literature Review: Optimization under Uncertainty

Dependent on objectives, constraints, and decision variables, the literature on deterministic programming models categorizes problems as linear programming [10], non-linear programming [11], and integer programming [12], among others. However, real-life data are usually not certain, and some methods have been proposed for treating such parameter uncertainty. One conventional approach is sensitivity analysis [13], which deals with uncertainty after finding the optimal solution.

Other frameworks that explicitly incorporate uncertainty into the computation of the optimal solution are stochastic programming, dynamic programming, and robust optimization [14]. Although the above-mentioned methods overlap, they have unfolded independently of each other. Stochastic programming incorporates stochastic components into the programming framework. The method represents uncertain data by scenarios via for example, Monte Carlo sampling, and simple average approximation. Dynamic programming deals with stochastic uncertain systems in a multi-stage framework. It is a technique more widely utilized in derivative pricing as it tackles problems with uncertain coefficients over multiple horizons. In recent times, robust optimization method is a widely acceptable approach in tackling uncertainty. Robust optimization models uncertainty by using a certain membership (uncertainty sets that are based on statistical estimates and probabilistic guarantees on the solution) and optimizes the worst possible case of the problem. When the uncertain parameters are known within certain bounds, robust optimization is best suited [14].

Let us consider a general stochastic programming problem:

$$\text{maximize} \quad f(x) = \text{maximize} \quad \mathbb{E}[F(x,\zeta)], \tag{3}$$

where the expectation is taken over ζ. Here, the objective function $F(x,\zeta)$ is dependent on decision variable x and uncertain parameter ζ. The objective function is well defined, as it is optimized on the average. An important question often asked is what if the uncertainty resides in the constraints [15]. One approach is to formulate such problems similarly by incorporating penalties for constraint violations [9]. An alternative approach also employed in this study is to require that the constraints are satisfied for all possible values of the uncertain parameters with a high probability. Contemporary work in robust optimization has resulted in defining and specifying uncertainty sets to guarantee that chance constraints are satisfied with a targeted probability, thus providing a connection between stochastic programming and robust optimization. For the purpose of this study, the theory of chance-constrained models is explored further.

The chance-constrained stochastic optimization method is one of the major approaches to solving optimization problems under uncertainty. It ensures that an individual constraint is satisfied with a target probability. Mainly, it restricts the feasible region so that a solution is obtained at a high probability. Chance-constrained programming was first investigated by [16] to ensure that the optimal solution satisfied constraints at a certain probability or confidence level. Many research works have now delved into more ways of tackling chance-constrained problems and increasing the efficiencies of such optimization problems.

A general chance-constrained programming problem takes the form:

$$\text{maximize} \quad f(x) \tag{4}$$
$$\text{subject to} \quad \mathbb{P}\{g(x,\zeta) \leq 0\} \geq \alpha, \tag{5}$$
$$x \in \Lambda, \tag{6}$$

where x denotes decision variables, Λ denotes a set of all feasible solutions, ζ represents uncertain parameters, and $\alpha \in (0,1)$ is a desired safety factor chosen by the modeler. The chance constraint ensures that the constraint $g(x,\zeta) \leq 0$ is satisfied with a probability α at least.

Chance-constrained optimization problems are challenging computationally. Even checking the feasibility of a chance constraint is NP-hard, and the feasible region is usually non-convex. It is also difficult to obtain samples to estimate the uncertain parameter's probability distribution accurately. In practice, assumptions about the probability distribution of the uncertain parameters in a chance-constrained problem need to be made to express the probabilistic constraint (5) in closed form. It is, however, difficult to obtain an equivalent deterministic constraint for most probability constraints.

A more difficult challenge that arises is the need to commit to a particular distribution of the uncertain parameter ζ given only restricted information about the stochastic parameters [5]. To avoid the above difficulties such as a selection of the proper probability distribution of the uncertain parameter, NP-hard feasibility checking, and nonconvexity, approximation methods have been proposed. In general, there exist two kinds of approximation approaches for a chance constraint: the analytical approximation method and the sampling-based method. Given the disadvantages of the sampling-based method such as the use of an empirical distribution of the random samples to model the actual distribution [17] among others, we pursue the analytical approximation approach. The analytical approximation method formulates the chance constraint into an equivalent deterministic counterpart. Robust optimization presents a way to approximate analytically a chance constraint. This technique requires a mild assumption on the probability distribution of the uncertain parameters and provides a tractable and feasible solution to the chance-constrained problem. Research contributions using the framework of robust counterpart optimization were explored by [18,19].

A natural way to tackle a chance constraint against parameter uncertainty, which is itself characterized by an uncertain probability distribution, is to use the Distributionally-Robust Chance-Constrained (DRCC) approach, a variant of distributionally-robust optimization. Distributionally-robust optimization is an approach that bridges the gap between robust optimization and stochastic programming [20]. In particular, the distributionally-robust chance constraint can be expressed as:

$$\inf_{\mathbb{P} \in \mathcal{P}} \mathbb{P}\{g(x,\zeta) \leq 0\} \geq \alpha, \tag{7}$$

where $\mathcal{P}$ represents a set of all probability distributions that is in line with the characteristic properties of the true probability $\mathcal{D}$ such as moment information or its support [21]. Whenever x satisfies (7) and $\mathcal{D} \in \mathcal{P}$ is the true distribution, x satisfies the chance constraint (5) under the true probability distribution $\mathcal{D}$.

In the DRCC paradigm, the distribution of ζ is not exactly known, but rather assumed to belong to a given set $\mathcal{P}$. In other words, Equation (7) requires that for all probability distributions of ζ, the chance constraint holds. In the DRCC framework, the work in [22] investigated safe tractable approximations of chance-constrained affinely-perturbed linear matrix inequalities. The work in [7] showed that in some cases of a linear chance constraint problem, the worst case moment expression could be analytically expressed. Based on S-lemma, the work in [23] showed that a distributionally-robust chance constraint is tractable when $f(x,\zeta)$ is linear in the decision variable x and piecewise linear or quadratic in the uncertainty parameter ζ. In this study, to obtain a well-posed optimization problem without assuming full knowledge of the probability measure, in moment-based optimization, a distributionally-robust counterpart to a defined chance constraint of capital requirement is considered to guarantee satisfying bank capital requirements.

3. Problem Definition and Assumption: Chance Constraint with an Unknown Distribution

The proposed asset-liability optimization model is based on constraint robustness with the chance constraint of capital to risk assets ratio in a safety-first framework under the condition that only moment information is known. This paper aims to extend the proposed single-objective capital to the risk asset ratio chance-constrained optimization model in [4] by considering the multi-objective constraint robustness approach in a modified safety-first framework.

The following assumptions were made to develop the model: First, the set of all probability distributions have known first and second moments. Second, the set of probability distributions were assumed to include the true probability distribution of the random vector.

Motivated by [7], this study examines an aspect of a chance-constrained robust problem with known first and second moments. For the purpose of completeness, a reintroduction of the description of the notations and parameters is needed.

Notation and Parameter Description

Without loss of generality, define the single generic constraint as:

$$\mathbb{P}\left\{y^0(x) + y(x)^T \zeta \leq 0\right\} \geq \alpha, \ \alpha \in (0,1) \tag{8}$$

and define the random vector:

$$r = \begin{bmatrix} 1 & \zeta^T \end{bmatrix}^T \in \mathbb{R}^{h+1}$$

and:

$$\hat{r} = E\{r\} = E\{\begin{bmatrix} 1 & \zeta^T \end{bmatrix}^T\} = \begin{bmatrix} 1 & \hat{\zeta}^T \end{bmatrix}^T$$
$$\Gamma = var\{r\} = var\{\begin{bmatrix} 1 & \zeta^T \end{bmatrix}\}$$

Consider $v \leq h+1$ as the rank of Γ and $\Gamma_{f.r} \in \mathbb{R}^{h+1}$ as a full rank factor such that $\Gamma = \Gamma_{f.r}\Gamma_{f.r}^T$.

Let:
$$\tilde{z} = [y^0(x) \ \ y(x)^T]^T \in \mathbb{R}^{h+1}$$

Let us define the quantity:
$$\varphi(z) = r^T \tilde{z}$$

and:

$$\hat{\varphi}(z) = E\{\varphi(z)\} = \hat{r}^T \tilde{z} \tag{9}$$
$$\sigma^2(z) = var\{\varphi(z)\} = \tilde{z}^T \Gamma \tilde{z} \tag{10}$$

The normalized random variable is defined as:

$$\tilde{\varphi}(z) = \frac{\varphi(z) - \hat{\varphi}(z)}{\sigma(z)}$$

Therefore, one can rewrite Constraint (1) as:

$$\mathbb{P}\{r^T \tilde{z} \leq 0\} = \mathbb{P}\{\varphi(z) \leq 0\} = \mathbb{P}\{\tilde{\varphi}(z) \leq -\hat{\varphi}(z)/\sigma(z)\} \geq \alpha.$$

By distributional constraint robustness, the chance constraint $\mathbb{P}\{r^T \tilde{z} \leq 0\} \geq \alpha$ should be robustly enforced by considering the problem:

$$\inf_{r \sim \mathcal{P}} \mathbb{P}\{r^T \tilde{z} \leq 0\} \geq \alpha = \inf_{r \sim (\hat{r}, \Gamma)} \mathbb{P}\{r^T \tilde{z} \leq 0\} \geq \alpha \tag{11}$$

where $r \sim \mathcal{P}$ means that the distribution of r belongs to the family $\mathcal{P}$ with $\mathcal{P}$ having known first and second moments.

4. Model Formulation and Approach

4.1. Formulation

Let us consider the random variable $\mathbf{Y}$ such that $\mathbf{Y}^T \tilde{z} = (2\hat{r}^T - r^T)\tilde{z}$. From the result of [24] on the tight bound Chebyshev inequality [25], the following holds:

$$\sup_{r \sim (\hat{r}, \Gamma)} \mathbb{P}\{r^T \tilde{z} > 0\} = \sup_{r \sim (\hat{r}, \Gamma)} \mathbb{P}\{\mathbf{Y}^T \tilde{z} - \hat{r}^T \tilde{z} > \hat{r}^T \tilde{z}\}$$

$$\leq \begin{cases} \dfrac{1}{1 + \frac{(\hat{r}^T \tilde{z})^2}{\tilde{z}^T \Gamma \tilde{z}}}, & \text{if } \hat{r}^T \tilde{z} \geq 0 \\ 1, & \text{Otherwise.} \end{cases} \tag{12}$$

Obviously,

$$\mathbb{P}\{\mathbf{Y}^T \tilde{z} - \hat{r}^T \tilde{z} > \hat{r}^T \tilde{z}\} = \mathbb{P}\{\hat{r}^T \tilde{z} - \mathbf{Y}^T \tilde{z} < -\hat{r}^T \tilde{z}\}$$
$$= \mathbb{P}\{r^T \tilde{z} - \hat{r}^T \tilde{z} < -\hat{r}^T \tilde{z}\}.$$

This, combined with (12), implies that:

$$\sup_{r \sim (\hat{r}, \Gamma)} \mathbb{P}\{r^T \tilde{z} - \hat{r}^T \tilde{z} < -\hat{r}^T \tilde{z}\} \leq \frac{\tilde{z}^T \Gamma \tilde{z}}{\tilde{z}^T \Gamma \tilde{z} + (\hat{r}^T \tilde{z})^2}$$

Hence,

$$\frac{\tilde{z}^T \Gamma \tilde{z}}{\tilde{z}^T \Gamma \tilde{z} + (\hat{r}^T \tilde{z})^2} \leq 1 - \alpha$$

is sufficient for Constraint (11) to hold. The expression can be recast in various forms as:

$$\tilde{z}^T \Gamma \tilde{z} \leq (1 - \alpha)(\tilde{z}^T \Gamma \tilde{z} + (\hat{r}^T \tilde{z})^2),$$

$$\alpha \tilde{z}^T \Gamma \tilde{z} \leq (1 - \alpha)(\hat{r}^T \tilde{z})^2,$$

$$\frac{\alpha}{(1 - \alpha)} \tilde{z}^T \Gamma \tilde{z} \leq (\hat{r}^T \tilde{z})^2,$$

$$(\hat{r}^T \tilde{z})^2 \geq \tilde{z}^T \Gamma \tilde{z} \frac{\alpha}{(1 - \alpha)},$$

$$\hat{\varphi}^2(z) \geq \sigma^2(z) \frac{\alpha}{(1 - \alpha)}.$$

Theorem 1. *The chance constraint, $\mathbb{P}\{\hat{r}^T\tilde{z} \leq 0\} \geq \alpha$, for any $\alpha \in (0,1)$ expressed as a constraint robust term in the form $\inf_{r\sim(\hat{r},\Gamma)} \mathbb{P}\{r^T\tilde{z} \leq 0\} \geq \alpha$ is equivalent to the second-order cone constraint $(\hat{r}^T\tilde{z})^2 \geq \tilde{z}^T\Gamma\tilde{z}\frac{\alpha}{(1-\alpha)}$.*

4.2. CreditMetrics Approach

Future bank capital depends on future market values of assets and liabilities. This paper employs a modified CreditMetrics approach to estimate the future market value of loans, i.e., the uncertainty parameter in (1). A modified CreditMetrics approach has been proposed by [4]. We adopted this approach for the sole aim of determining the future market value of loans and the associated risk measure.

Migration of Ratings

Credit rating transition is the migration of loans across different ratings over a risk period. Credit risk arises from changes in the loan value as a result of upgrades and downgrades. Therefore, it is prudent to evaluate the probability of default and the possibility of migration to other ratings. Table 1 shows a transition matrix developed from historical data by CRISIL (Credit Rating Information Services of India Limited). The likelihood that an A borrower will remain at A over the next year is 71.23%.

Table 1. CRISIL's one-year mean transition rates (1993–2014)(%). Source: CRISIL Default Study 2014.

Rating	AAA	AA	A	BBB	BB	B	C	D
AAA	98.23	1.54	0.23	0.00	0.00	0.00	0.00	0.00
AA	17.04	78.52	3.70	0.74	0.00	0.00	0.00	0.00
A	9.59	15.07	71.23	4.11	0.00	0.00	0.00	0.00
BBB	4.02	3.29	15.69	76.28	0.00	0.37	0.37	0.00
BB	11.11	22.22	22.22	22.22	22.22	0.00	0.00	0.00
B	0.00	0.00	0.00	0.00	0.00	50.00	0.00	50.00
C	0.00	0.00	0.00	0.00	0.00	0.00	0.00	100

The recovery rate is a measure of the extent to which a creditor recovers the principal and accrued interest due on a defaulted debt [26]. According to Moody's "Default and Recovery rates for project bank loans, 1983–2014", the ultimate recovery rates average 80%, which is roughly consistent with existing works of [27] (84.14%), [28] (81.12%), [29] (80%), and [30] (87%). For the purpose of this study, we consider loan recovery rates estimated by Moody's Investors Service.

4.3. Loan Valuation and Credit Risk

In this section, the year-ahead market value of loans and credit risk are estimated. Reference is made to the approach employed by [4] and followed accordingly. Table 2 contains entries of $(j-1)$-year forward rates, $f_{cr}^{1,j}$, which are spot rates from now with credit rating cr.

Table 2. One-year forward zero curve for each credit rating category (%). Adapted from [4] with permission from authors.

Category	Year 1 ($f_{cr}^{1,2}$)	Year 2 ($f_{cr}^{1,3}$)	Year 3 ($f_{cr}^{1,4}$)	Year 4 ($f_{cr}^{1,5}$)
AAA	3.60	4.17	4.73	5.12
AA	3.65	4.22	4.78	5.17
A	3.72	4.32	4.93	5.32
BBB	4.10	4.67	5.25	5.63
BB	5.55	6.02	6.78	7.27
B	6.05	7.02	8.03	8.52
CCC/C	15.05	15.02	14.03	13.52

Please refer to [4] for the mathematical expression of the forward value of a unit capital invested in the kth loan with either a default (ζ_{kdt}) or non-default (ζ_{kt}) migration path, expected forward loan value ($E(\zeta_k)$), and variance of unit capital invested ($Var(\zeta_k)$).

Credit risk arises because loan values can vary depending on the credit quality changes, and so, any reasonable risk measure must reflect this variability. CreditMetrics proposed two measures to characterize credit risk: variance or standard deviation and percentile level [31,32]. The risk measures reflect the portfolio distribution, and both contribute in the effort to quantify risk. The two risk measures also reflect potential losses from the same portfolio distribution. The variance measure reflects how different the expected value of the loans will be from the actual value, and the percentile levels aid in arriving at the unexpected losses of the portfolio.

The specified percentile level is interpreted as the lowest value the loan portfolio will achieve for say 5% of the time: the fifth percentile. Therefore, the likelihood that the true loan portfolio value is less than the calculated fifth percentile level is only 5%. Given the full distribution of loan portfolio values, ζ_{kt} and ζ_{kdt}, one can derive the percentile level. It is important to note that the two credit risk measures (variance and the use of percentile level) give different values and must be interpreted in a different manner. According to the CreditMetrics Technical document [32], the process of estimating credit risk via percentile levels, for e.g. fifth percentile, is as follows: First, the fifth-percentile level number is obtained from the loan value distribution, and second, using the fifth percentile, the amount of credit risk is estimated by taking the difference of the mean portfolio and the fifth percentile level number.

The percentile level approaches often used in the literature are Value-at-Risk (VaR) and Conditional Value-at-Risk (CVaR), as they assess risk within the full context of a portfolio. The percentile level is naturally appealing to employ as it shows precisely the likelihood that the portfolio value will fall below that number.

This study in the context of credit risk infers VaR and CVaR from a distribution of the value of the portfolio and not from a distribution of the losses in the portfolio. Credit VaR is defined as the distance from the percentile to the mean of the forward distribution, at the desired confidence level. Credit CVaR is the average distance beyond VaR from the percentile to the mean of the forward distribution, at the desired confidence level. Both reflect (average) unexpected credit loss at the desired confidence level. However, there is another way to look at credit VaR if one considers the distribution of losses: it can also be interpreted as percentile loss itself, i.e., including expected losses.

This paper uses the credit VaR and credit CVaR to replace the market VaR and market CVaR used in a modified safety-first framework described in [9] to avert extreme unexpected credit losses and control the downside risk.

5. Model Development

The asset-liability optimization model based on the chance constraint of capital to the risk (weighted) asset requirement is structured and presented in this section. Objective functions consider maximizing the annual interest rate and expected portfolio value, minimizing credit risk, and providing a safety-net against extreme losses.

Given definitions of input parameters and replacing chance Constraint (1) with its robust constraint counterpart (2) provide the following distributionally-robust chance-constrained problem:

$$\begin{aligned}
&\underset{x}{\text{maximize}} && R(x) \\
&\underset{x}{\text{maximize}} && \mu(x) \\
&\underset{x}{\text{minimize}} && \sigma^2(x) \\
&\underset{x}{\text{minimize}} && \frac{(1-\alpha)\text{creditCVaR}(x)}{\text{creditVaR}(x) - R} \\
&\text{subject to} && \underset{\mathbb{P}\in\mathcal{P}}{\inf}\ \mathbb{P}\{y^0(x) + y(x)^T \zeta \le 0\} \ge \alpha \\
& && \sum_{k=1}^{u+v} x_k = 1 \\
& && x_1, x_2, \ldots, x_{u+v} \ge 0
\end{aligned} \qquad (13)$$

Based on Theorem 1, the robust chance constraint is equivalent to a deterministic convex counterpart. Therefore, the multi-objective robust chance-constrained optimization model is:

$$\begin{aligned}
&\underset{x}{\text{maximize}} && R(x) \\
&\underset{x}{\text{maximize}} && \mu(x) \\
&\underset{x}{\text{minimize}} && \sigma^2(x) \\
&\underset{x}{\text{minimize}} && \frac{(1-\alpha)\text{creditCVaR}(x)}{\text{creditVaR}(x) - R} \\
&\text{subject to} && \hat{\varphi}^2(z) \ge \sigma^2(z)\frac{\alpha}{(1-\alpha)} \\
& && \sum_{k=1}^{u+v} x_k = 1 \\
& && x_1, x_2, \ldots, x_{u+v} \ge 0
\end{aligned} \qquad (14)$$

where $\alpha \in (0,1)$, $\hat{\varphi}(z) = \hat{r}^T \tilde{z}$, $\sigma(z) = \tilde{z}\Gamma\tilde{z}$, and $\tilde{z} = [y^0(x)\ y(x)^T]^T$.

Given a portfolio x for u number of loans, the probability of the loss function not exceeding an acceptable threshold δ is:

$$\Xi(x,\delta) = \int_{f(x,\zeta) \le \delta} p(\zeta)d\zeta. \qquad (15)$$

For a confidence level α, the VaR of portfolio x is given by:

$$\text{VaR}_\alpha(x) = \min\{\delta \mid \Xi(x,\delta) \ge \alpha\}. \qquad (16)$$

The corresponding CVaR is expressed as conditional expectation of the loss of the portfolio exceeding or equal to VaR, i.e., when all random values are continuous, the following is derived:

$$\begin{aligned}
\text{CVaR}_\alpha(x) &= \mathbb{E}\{f(x,\zeta) \mid f(x,\zeta) \ge \text{VaR}_\alpha(x)\} \\
&= \frac{1}{1-\alpha}\int_{f(x,\zeta) \ge \text{VaR}_\alpha(x)} f(x,\zeta)p(\zeta)d\zeta.
\end{aligned} \qquad (17)$$

The work in [33] proposed an equivalent function for CVaR. They expressed their idea as:

$$\text{CVaR}_\alpha(x) = \min F_\alpha(x,\delta), \qquad (18)$$

where:
$$F_\alpha(x,\delta) = \delta + \frac{1}{1-\alpha} \int_{f(x,\zeta) \geq \delta} (f(x,\zeta) - \delta) p(\zeta) d\zeta.$$

One can find the optimal CVaR by solving the right-hand side of Equation (18). In order to minimize CVaR over x, we minimized the auxiliary function with respect to x and δ:

$$min_x CVaR_\alpha(x) = min_{x,\delta} F_\alpha(x,\delta). \tag{19}$$

The work in [33] presented an approximation to the auxiliary function $F_\alpha(x,\delta)$ via the sampling method:

$$\hat{F}_\alpha(x,\delta) = \delta + \frac{1}{(1-\alpha)u} \sum_{i=1}^{u} max(f(x,\zeta) - \delta, 0). \tag{20}$$

Comparing Equation (20) to Equation (18), the problem $min_x CVaR_\alpha(x)$ can be approximated by replacing $F_\alpha(x,\delta)$ with $\hat{F}_\alpha(x,\delta)$ in Equation (19):

$$\min_{x,\delta} \delta + \frac{1}{(1-\alpha)u} \sum_{i=1}^{u} max(f(x,\zeta) - \delta, 0). \tag{21}$$

To solve this optimization problem, one can replace $max(f(x,\zeta) - \delta, 0)$ with artificial variables z_i and impose constraints $z_i \geq f(x,\zeta) - \delta$ and $z_i \geq 0$:

$$\begin{aligned}
\min_{x,z,\delta} \quad & \delta + \frac{1}{(1-\alpha)u} \sum_{i=1}^{u} z_i \\
\text{subject to} \quad & z_i \geq 0, \quad i = 1, \ldots, u, \\
& z_i \geq f(x_i, \zeta_i) - \delta, \quad i = 1, \ldots, u, \\
& x \in \Lambda.
\end{aligned} \tag{22}$$

For a portfolio of u loans, we assumed $\mu_i \in \mathbb{R}^u$ is the random vector of the expected returns value of ζ with a probability density function $p(\mu)$. To determine the mean loss of the portfolio, this study defines the loss function as $f(x_i, \mu_i) = -\sum_{i=1}^{u} x_i \mu_i = -[\mu_1 x_1 + \ldots + \mu_u x_u]$. Since the loss function is convex, then the auxiliary function $F_\alpha(x,\delta)$ is a also a convex function and can be solved using well-known optimization techniques.

Representing each objective function by $P_1(x), \ldots, P_4(x)$, the multi-objective robust chance-constrained optimization problem (23) can be approximated via the approximation technique (22) as shown below:

$$\begin{aligned}
\underset{x}{\text{maximize}} \quad & P_1(x) = \mathbf{R}(x) \\
\underset{x}{\text{maximize}} \quad & P_2(x) = \mu(x) \\
\underset{x}{\text{minimize}} \quad & P_3(x) = \sigma^2(x) \\
\underset{x}{\text{minimize}} \quad & P_4(x) = \frac{(1-\alpha)(\delta + \frac{1}{(1-\alpha)u} \sum_{i=1}^{u} z_i)}{\delta - R} \\
\text{subject to} \quad & \hat{\varphi}^2(z) \geq \sigma^2(z) \frac{\alpha}{(1-\alpha)}, \\
& \delta \geq R, \\
& z_i \geq 0, \quad i = 1, \ldots, u, \\
& z_i + x_i^T \mu_i + \delta \geq 0, \quad i = 1, \ldots, u, \\
& \sum_{k=1}^{u+v} x_k = 1, \\
& x_1, x_2, \ldots, x_{u+v} \geq 0,
\end{aligned} \qquad (23)$$

where $\alpha \in (0,1)$, $\hat{\varphi}(z) = \hat{r}^T \tilde{z}$, $\sigma^2(z) = \tilde{z} \Gamma \tilde{z}$, and $\tilde{z} = [y^0(x) \ y(x)^T]^T$.

Transformation and Solution to the Multi-objective Model

The multi-objective portfolio optimization model (23) addresses the trade-off between conflicting or competing objectives. This type of problem is referred to as a Polynomial Goal Programming (PGP) problem. The idea behind such an approach is to obtain smaller computable elements of the problem and then find solutions iteratively that meet individual goals.

Generally, there will not be a single solution of Problem (23) that can maximize both Objective 1 and Objective 2. Alternately, the solution of the multi-objective optimization problem (23) has to be obtained in a two-step process. First, individually solve each objective $P_1(x), P_2(x), P_3(x)$, and $P_4(x)$ subject to the constraints. Denote optimal values (desired goals) after solving the individual objectives subject to constraints by P_1^*, P_2^*, P_3^*, and P_4^*. Second, find an optimal solution that preserves individual objectives by minimizing the deviation of each individual objective from the ideal solution. Let d_1, d_2, d_3, and d_4 be non-negative variables that account for deviation from the desired goals, P_1^*, P_2^*, P_3^*, and P_4^*. The multi-objective optimization problem (23) transforms into a single objective problem in a specific form of the general Minkowski distance defined as $O = \left\{ \sum_{f=1}^{h} \left| \frac{d_f^p}{P_f} \right| \right\}^{1/p}$, where P_f represents the corresponding desired goal and is used as basis for normalization of the fth variable.

The single objective optimization problem derived via PGP for the multi-objective problem (23) is defined as:

$$\underset{x,d}{\text{minimize}} \quad O(x) = |\frac{d_1}{P_1^*}| + |\frac{d_2}{P_2^*}| + |\frac{d_3}{P_3^*}| + |\frac{d_4}{P_4^*}|$$

$$\text{subject to} \quad R(x) + d_1 = P_1^*$$

$$\mu(x) + d_2 = P_2^*$$

$$\sigma^2(x) - d_3 = P_3^*$$

$$\frac{(1-\alpha)(\delta + \frac{1}{(1-\alpha)u}\sum_{i=1}^{u} z_i)}{\delta - R} - d_4 = P_4^*$$

$$(\hat{r}^T \tilde{z})^2 \geq \tilde{z} \Gamma \tilde{z} \frac{\alpha}{(1-\alpha)}, \tag{24}$$

$$\delta \geq R,$$

$$z_i \geq 0, \quad i = 1, \ldots, u,$$

$$z_i + x_i^T \mu_i + \delta \geq 0, \quad i = 1, \ldots, u,$$

$$\sum_{k=1}^{u+v} x_k = 1$$

$$d_f \geq 0, \quad f = 1, \ldots, 4$$

$$x_1, x_2, \ldots, x_{u+v} \geq 0$$

where $\alpha \in (0,1)$ and $\tilde{z} = [y^0(x) \quad y(x)^T]^T$.

Given that the objective function of Problem (24) has a fractional component, an alternative equivalent program can be deduced. Under the assumption that the feasible region is non-empty and bounded, the transformation:

$$w = \frac{1}{\delta - R}, \quad w \geq 0,$$

$$\tilde{x} = xw$$

translates (24) to the equivalent single objective optimization problem via PGP as:

$$\begin{aligned}
\underset{\tilde{x},d,w,z,\delta}{\text{minimize}} \quad & \left|\frac{d_1}{P_1^*}\right| + \left|\frac{d_2}{P_2^*}\right| + \left|\frac{d_3}{P_3^*}\right| + \left|\frac{d_4}{P_4^*}\right| \\
\text{subject to} \quad & \mathbf{R}^T \tilde{x} + d_1 = P_1^*, \\
& \mu^T \tilde{x} + d_2 = P_2^*, \\
& \tilde{x}^T Q \tilde{x} - d_3 = P_3^*, \\
& (w - \alpha w)\left(\delta + \frac{1}{(1-\alpha)\mathbf{T}} \sum_{i=1}^{\mathbf{T}} z_i\right) - d_4 = P_4^*, \\
& (\hat{r}^T \tilde{z})^2 \geq \tilde{z} \Gamma \tilde{z} \frac{\alpha}{(1-\alpha)}, \\
& \sum_{k=1}^{u+v} \tilde{x}_k = w, \\
& d_f \geq 0, \quad f = 1, \ldots, 4, \\
& \delta w - Rw = 1, \\
& w \geq 0, \\
& z_i w \geq 0, \quad i = 1, \ldots, u, \\
& z_i w + \tilde{x}_i^T \mu_i + \delta w \geq 0, \quad i = 1, \ldots, u, \\
& \tilde{x}_1, \tilde{x}_2, \ldots, \tilde{x}_{u+v} \geq 0,
\end{aligned} \quad (25)$$

where Q is covariance matrix of unit capital invested in loans, $\alpha \in (0,1)$, and $\tilde{z} = [y^0(x) \ y(x)^T]^T$. One can obtain an optimal solution to the problem by rescaling $\tilde{x}^*$ so that $x^* = \frac{\tilde{x}^*}{w^*}$.

6. Numerical Examples

In this section, the proposed model is subjected to numerical experiments [34] by considering a hypothetical bank operating in the U.S. Loan data are private information and difficult to obtain. This study defines and uses references and trusted sources such as World Bank, Moody's Investors Service, and CRISIL to back the data used for this section.

6.1. Data

Consider the asset structure of a hypothetical bank in the U.S. Let the bank's total loan and treasury bill amount, M, be $\$600,000$; the bank's total liability TL equal $\$1,192,000$, disclosed reserves DR of $\$166,000$, revaluation reserves Rr of $\$12,000$, and general loan loss provisions LP of $\$15,000$. The financial assets characterizing the financial environment on the bank are five types of loans, a treasury bill, fixed assets, and non-interest earning assets. Table 3 presents the information about the assets and capital funds to be allocated.

Interest rate ($\mathbf{R}$) for individual loans was set to lending interest rates for the U.S. quoted by the World Bank in 2014 as a reference. World Bank's one-year treasury bill rate of 0.1% for the U.S. in 2014 was used. The standardized approach of the Basel Accord was a reference point for risk weights. Moody's recovery rate of 80% was used for recovery rates for loan types. Let us consider correlations among loan borrowers as shown in Table 4. In financial crises, correlations of random assets tend to converge positively, maybe even to one. To represent a market in distress, we consider the correlation matrix in Table 4 to see the effects on capital adequacy.

Table 3. Asset structure of a U.S. Bank (IR, Interest Rate; RR, Recovery Rate; RW, Risk Weights; CV, Credit VaR; CCV, Credit CVaR).

Assets	Collateral	IR (%)	RR (%)	RW (%)	Mean	Variance	CV	CCV	Allocation
(ζ_1) 3-Year AAA Commercial and Industrial Loan	Inventory or account receivables	3.56	80.00	20	1.0243	4.1679×10^{-8}	0.2402	0.2829	0.2550
(ζ_2) 5-Year AA Agriculture and Farm Loan	Equipment, crops, livestock, etc.	4.17	80.00	50	1.0081	1.7196×10^{-5}	0.2558	0.2889	0.1663
(ζ_3) 2-Year BBBPersonal Loan	Savings account, tangible property, etc.	3.11	80.00	75	1.0213	4.2935×10^{-4}	0.2323	0.2140	0.2665
(ζ_4) 3-Year B Small Business Loan	Land, savings account, etc	4.03	80.00	75	0.81911	4.0625×10^{-3}	0.0258	0.0689	0.0468
(ζ_5) 4-Year A Auto Loan	Savings account or car itself	4.21	80.00	75	1.0248	4.7805×10^{-5}	0.2512	0.2892	0.2386
1-Year Treasury Bill	Not Applicable	0.1	100	0	1.001	0	0	0	0.0268
Fixed Assets	Not Applicable	0	100	0	1	0	0	0	-
Non-Interest Earning Assets	Not Applicable	0	100	0	1	0	0	0	-

Table 4. Asset correlations among loans of a market in crisis.

Risky Assets	ζ_1	ζ_2	ζ_3	ζ_4	ζ_5
ζ_1	1	0.85	0.8	0.8	0.8
ζ_2	0.85	1	0.9	0.85	0.8
ζ_3	0.8	0.9	1	0.9	0.8
ζ_4	0.8	0.85	0.9	1	0.95
ζ_5	0.8	0.8	0.8	0.95	1

6.1.1. Objective Function

The objective function is filled with various entries of Table 3 and some other computed results. Thus,

$$\begin{aligned}
&\underset{x}{\text{maximize}} && \mathbf{R}^T x \\
&\underset{x}{\text{maximize}} && \mu^T x \\
&\underset{x}{\text{minimize}} && x^T Q x \\
&\underset{x}{\text{minimize}} && \frac{(1-\alpha)\text{creditCVaR}(x)}{\text{creditVaR}(x) - R}, \quad \alpha = 0.95 \text{ and } R = 0.009.
\end{aligned} \tag{26}$$

6.1.2. CRAR Constraint

This study employed the CRAR developed by Basel III as the designated regulation requirement. From the definition of CRAR, i.e.,

$$\frac{M\Omega^T x - TL + DR + Rr + LP}{M(\varpi_1 \zeta_1 x_1 + \ldots + \varpi_u \zeta_u x_u + \varpi_{u+1} \tilde{\zeta}_{u+1} x_{u+1} + \ldots + \varpi_{u+v} \tilde{\zeta}_{u+v} x_{u+v})} \geq \lambda$$

where λ is the total capital requirement. The chance-constrained model based on the capital to risk assets ratio can be expressed as:

$$\mathbb{P}\left\{ y^0(x) + y(x)^T \tilde{\zeta} \leq 0 \right\} \geq \alpha \tag{27}$$

where:

$$y^0(x) = TL - DR - Rr - LP - M \sum_{k=u+1}^{u+v} (1 + \mathbf{R}_k) x_k$$

$$y(x) = M \sum_{k=1}^{u} (\lambda \varpi_k - 1) x_k$$

The year-ahead market value of net assets of the bank should be $600,000(\zeta^T x + \tilde{\zeta}_1 x_6) + 707,000 - 1,192,000 + 166,000 + 12,000 + 15,000$ where $600,000(\tilde{\zeta}_1 x_6) + 707,000$ represents the year-ahead market value of riskless assets. In this study, the total capital requirement ratio λ of 11% was used. The constraint based on CRAR is:

$$\underset{\mathbb{P} \in \mathcal{P}}{\inf} \ \mathbb{P}\left(\frac{600,000(\zeta^T x + \tilde{\zeta}_1 x_6) + 707,000 - 1,192,000 + 166,000 + 12,000 + 15,000}{1,500,000(0.2 \zeta_1 x_1 + 0.5 \zeta_2 x_2 + \ldots + 0.75 \zeta_5 x_5)} \geq 0.11 \right) \geq 0.95 \tag{28}$$

$$\underset{\mathbb{P} \in \mathcal{P}}{\inf} \ \mathbb{P}\left\{ y^0(x) + y(x)^T \tilde{\zeta} \leq 0 \right\} \geq 0.95 \tag{29}$$

where:

$$y^0(x) = 1,192,000 - 166,000 - 12,000 - 15,000 - 600,000(1 + 0.0010)x_6 - 707,000$$
$$y(x)^T = 600,000\left((0.11 \cdot 0.2 - 1)x_1 + (0.11 \cdot 0.5 - 1)x_2 + \ldots + (0.11 \cdot 0.75 - 1)x_5\right)$$

According to Theorem 1, Equation (28) is equivalent to:

$$(\hat{p}^T \tilde{z})^2 \geq \tilde{z}^T \Gamma \tilde{z} \left(\frac{0.95}{0.05} \right) \tag{30}$$

where:

$$\tilde{z} = \begin{bmatrix} 292000 - 604800x_6 \\ -586800x_1 \\ -567000x_2 \\ -550500x_3 \\ -550500x_4 \\ -550500x_5 \end{bmatrix}$$

$$\Gamma = \begin{bmatrix} 0 & 0 & 0 & 0 & 0 & 0 \\ 0 & 4.1679e-8 & 7.1146e-7 & 3.3807e-6 & 1.0403e-5 & 1.1269e-6 \\ 0 & 7.1146e-7 & 1.7196e-5 & 7.6383e-6 & 2.2199e-5 & 2.2632e-5 \\ 0 & 3.3807e-6 & 7.6383e-6 & 4.2935e-4 & 0.0012 & 1.1426e-4 \\ 0 & 1.0403e-5 & 2.2199e-5 & 0.0012 & 4.0625e-3 & 4.1755e-4 \\ 0 & 1.1269e-6 & 2.2632e-5 & 1.1426e-4 & 4.1753e-4 & 4.7805e-5 \end{bmatrix}$$

and:

$$\hat{r}^T = \begin{bmatrix} 1 & 1.0243 & 1.0081 & 1.0213 & 0.8911 & 1.0248 \end{bmatrix}$$

6.1.3. Constraint Based on Other Factors

The proportions of capital allocations must add up to one. Thus, the following holds.

$$\sum_{k=1}^{6} x_k = 1 \tag{31}$$

For diversification purposes, management sets up constraints with an upper bound limit of 0.4 with respect to loan allocations:

$$0 \leq x_1, \ldots, x_5 \leq 0.4 \tag{32}$$

Banks often allocate proportions to risky investment after considering riskless assets. For the purpose of this study, bank management allocates at most 5% of M to treasury bills.

$$0 \leq x_6 \leq 0.05 \tag{33}$$

The problem is transformed from a multi-objective optimization model to a single-objective model via Polynomial Goal Programming (PGP) as shown in (24). Thus,

$$\begin{aligned}
\underset{x,d}{\text{minimize}} \quad & \left|\frac{d_1}{P_1^*}\right| + \left|\frac{d_2}{P_2^*}\right| + \left|\frac{d_3}{P_3^*}\right| + \left|\frac{d_4}{P_4^*}\right| \\
\text{subject to} \quad & \mathbf{R}(x) + d_1 = P_1^* \\
& \mu(x) + d_2 = P_2^* \\
& \sigma^2(x) - d_3 = P_3^* \\
& \frac{(1-\alpha)(\delta + \frac{1}{(1-\alpha)u}\sum_{i=1}^{u} z_i)}{\delta - R} - d_4 = P_4^* \\
& (\hat{p}^T \tilde{z})^2 \geq \tilde{z} \Gamma \tilde{z} \frac{\alpha}{(1-\alpha)}, \\
& \delta \geq R, \\
& z_i \geq 0, \quad i = 1, \ldots, 5, \\
& z_i + x_i^T \mu_i + \delta \geq 0, \quad i = 1, \ldots, 5, \\
& \sum_{k=1}^{6} x_k = 1 \\
& d_f \geq 0, \quad f = 1, \ldots, 4 \\
& 0 \leq x_1, \ldots, x_5 \leq 0.4 \\
& 0 \leq x_6 \leq 0.05
\end{aligned} \quad (34)$$

where $\alpha = 0.95$, $R = 0.009$, and $\tilde{z} = [y^0(x) \quad y(x)^T]^T$. The above problem can be solved as shown in (25).

6.2. Results and Remarks

Mathematical computations were executed on a MacBook Pro (Intel(R) Core(TM) i7 @ 2.9 GHz, 16 GB RAM) with MATLAB (2017b).

Column 10 of Table 3 shows the optimal investment of the assets under consideration. The two-year BBB Personal Loan allocation value of 0.2665 is the highest amongst optimal investment proportions. This might be ascribed to its proper trade-off between the objective functions of our model. The high-risk weights are complimented by its high interest rates. The optimal solution minimizes the deviation of each objective from its ideal solution. The distributionally-robust CRAR chance constraint optimization model meets the Basel III Tier 1 capital requirements ratio of equal or more than 6% and also meets the total capital requirements of 11% considered for this study under the guidance of the Basel III capital requirements.

A sensitivity analysis was performed to determine the robustness of the proposed model by computing the CRAR value using loan values under the worst-credit migration path from Table 5. This was done with the dual aim of testing our model under the worst-case scenario, i.e., default and also to test model robustness. The value of CRAR even under the worst-case scenario confirmed that the bank was guaranteed to meet our Basel III total capital requirement of 11%.

To further investigate the results of our distributionally-robust CRAR optimization model, this study explored the CRAR chance constraint with the utilization of the optimal investment proportions and worst credit migration path values of loans. The expected values of the year-ahead market value of loans under the worst-case scenario were assigned to ζ, the optimal allocation proportions to x, and the treasury bill to ζ_1.

Table 5. Year-ahead or forward market value of loans under default.

Loan Type	Worst-Credit Migration Path	Value
3-Year AAA Commercial and Industrial Loan	$AAA \to CCC \to CCC \to D$	0.6712
5-Year AA Agriculture and Farm Loan	$AA \to CCC \to CCC \to CCC \to CCC \to D$	0.6193
2-Year BBB Personal Loan	$BBB \to CCC \to D$	0.7264
3-Year B Small Business Loan	$B \to CCC \to CCC \to D$	0.6800
4-Year A Auto Loan	$AAA \to CCC \to CCC \to CCC \to D$	0.6501

The findings of this paper explicitly disclose that in a safety-first framework, CRAR values will be greater than the Basel III minimum capital requirement at a confidence level of 95% if the model introduced is employed even if the model introduced is employed under the worst-case scenario. Therefore, the CRAR optimization model does guarantee that banks will cope with the capital requirements of Basel III with a greater likelihood of 95% irrespective of changes in the forward market value of assets. This approach also provides a safety-net against extreme losses and controls credit and capital risk.

6.2.1. Additional Remarks

To further explore our findings, this study subjected our robust approach for comparison to a stochastic model. This research used the same dataset used by [4].

We replaced the equivalent of our robust chance-constraint $((\hat{r}^T \tilde{z})^2 \geq \tilde{z} \Gamma \tilde{z} \frac{\alpha}{(1-\alpha)})$ in (34) with an equivalent term of the stochastic version $(F_{RG}^{-1}(\alpha)\sqrt{\tilde{z}\Gamma\tilde{z}} + \hat{r}^T \tilde{z} \leq 0)$ derived from Theorem 4.1 of [4]. $F_{RG}^{-1}(\alpha)$ was inverse for the standard right truncated Gaussian distribution cumulative probability function. All other parameters had the same definition in this paper as the following Table 6.

Table 6. Performance index of distributionally- and non-distributionally-robust models. CRAR, Capital to Risk Assets Ratio (CRAR).

	Distributionally Robust	Non-distributionally Robust
CRAR	12.2%	11.08%
Worst-case CRAR	8.9%	8.2%
Credit risk (variance)	0.0204	0.0203
Interest rate return	0.0565	0.0546
Portfolio value return	0.7661	0.8559

Under both CRAR chance constrained models, a Tier 1 capital ratio of equal or more than 6%, and the total capital ratio, i.e., CRAR equal to or greater than 11%, considered for this study were met. However, the robust chance constraint model meets the capital adequacy to a higher degree. Using loan values under the worst-credit migration path, the CRAR value of 8.9% was reported for the robust CRAR chance-constrained optimization model. The stochastic CRAR chance-constrained model reported a CRAR value of 8.2%. The managerial implications of the findings of this study are that the distributionally-robust model when applied will meet the capital requirements at a higher rate, maximize the interest rate return at the expense of a smaller portfolio value return while providing a better framework for treating parameter uncertainty.

7. Conclusions

This paper studied the asset-liability model, which is an extension of the single-objective CRAR chance-constrained model [4], by considering the multi-objective constraint robustness approach in a modified safety-first framework. Specifically, this study constructed a distributionally-robust optimization model in a safety-first framework under the capital to risk assets ratio chance constraint with uncertainty set based on the fact that only expectation and second marginal moment information were known. This approach, which is key to practical implications, on the one hand, provides banks with the guarantee of meeting capital regulation with a great probability of 95%, not only controlling the credit risk but also the capital risk. On the other hand, the approach provides a safety-net against extreme losses.

The proposed distributionally-robust capital to risk asset ratio chance-constrained optimization model guarantees banks will meet the capital requirements of Basel III with a likelihood of 95% irrespective of changes in the future market value of assets. A sensitivity analysis was performed to determine the robustness of the proposed model by computing the CRAR value using loan values under the worst-credit migration path. Even under the worst-case scenario, i.e., when loans default, our proposed capital to risk asset ratio chance-constrained optimization model meets the minimum total requirements of Basel III. The findings of this research are crucial for practitioners as they showcase a coherent manner to aid banks in meeting capital requirements.

Author Contributions: Conceptualization, E.F.E.A.M.; Data Curation, E.F.E.A.M. and B.Y.; Formal analysis, E.F.E.A.M., B.Y. and K.Z.; Funding acquisition, K.Z.; Investigation, E.F.E.A.M., B.Y. and K.Z.; Methodology, E.F.E.A.M., B.Y. and K.Z.; Software, E.F.E.A.M. and K.Z.; Validation, E.F.E.A.M., B.Y. and K.Z.; Visualization, E.F.E.A.M. and K.Z.; Supervision, B.Y.

Funding: Ebenezer Fiifi Emire Atta Mills and Kailin Zeng acknowledge support from School of Economics & Management, Jiangxi University of Science & Technology, Ganzhou, China, and Ganzhou Academy of Financial Research (GAFR), Ganzhou, China. Bo Yu acknowledges support by the National Nature Science Foundation of China (11571061).

Acknowledgments: We thank Mavis Agyapomah Baafi for her valuable suggestions. We also thank the anonymous reviewers for taking time off their busy schedule to review this paper.

Conflicts of Interest: The authors declare no conflict of interest.

References

1. Bernanke, B.S. Four Questions about the Financial Crisis. Available online: https://www.federalreserve.gov/newsevents/speech/bernanke20090414a.htm (accessed on 29 June 2019).
2. Fund, I.M. World Economic Outlook: Uneven Growth—Short and Long-Term Factors. 2015. Available online: https://www.imf.org/external/pubs/ft/weo/2015/01/ (accessed on 29 April 2019).
3. Hasan, I.; Siddique, A.; Sun, X. Monitoring the "invisible" hand of market discipline: Capital adequacy revisited. *J. Bank Financ.* **2015**, *50*, 475–492. [CrossRef]
4. Atta Mills, E.F.E.A.; Yu, B.; Gu, L. On meeting capital requirements with a chance-constrained optimization model. *SpringerPlus* **2016**, *5*, 500. [CrossRef] [PubMed]
5. Delage, E.; Ye, Y. Distributionally robust optimization under moment uncertainty with application to data-driven problems. *Oper. Res.* **2010**, *58*, 595–612. [CrossRef]
6. Scarf, H.; Arrow, K.; Karlin, S. A min-max solution of an inventory problem. In *Studies in the Mathematical Theory of Inventory and Production*; Stanford University Press: Redwood City, CA, USA, 1958; pp. 201–209.
7. Calafiore, G.C.; El Ghaoui, L. On distributionally robust chance-constrained linear programs. *J. Optim. Theory Appl.* **2006**, *130*, 1–22. [CrossRef]
8. Wiesemann, W.; Kuhn, D.; Sim, M. Distributionally robust convex optimization. *Oper. Res.* **2014**, *62*, 1358–1376. [CrossRef]
9. Atta Mills, E.F.E.A.; Yan, D.; Yu, B.; Wei, X. Research on regularized mean-variance portfolio selection strategy with modified Roy safety-first principle. *SpringerPlus* **2016**, *5*, 919. [CrossRef] [PubMed]
10. Dantzig, G. *Linear Programming and Extensions*; Princeton University Press: Princeton, NJ, USA, 2016. Available online: https://science.sciencemag.org/content/146/3651/1572 (accessed on 1 June 2019).
11. Kuhn, H.W.; Tucker, A.W. Nonlinear programming. In *Traces and Emergence of Nonlinear Programming*; Kjeldsen, T., Ed.; Birkhäuser: Basel, Switzerland, 2014; pp. 247–258.
12. Parreño-Torres, C.; Alvarez-Valdes, R.; Ruiz, R. Integer programming models for the pre-marshalling problem. *Eur. J. Oper. Res.* **2019**, *274*, 142–154. [CrossRef]
13. Saltelli, A.; Aleksankina, K.; Becker, W.; Fennell, P.; Ferretti, F.; Holst, N.; Li, S.; Wu, Q. Why so many published sensitivity analyses are false: A systematic review of sensitivity analysis practices. *Environ. Model. Softw.* **2019**, *114*, 29–39. [CrossRef]

14. Bertsimas, D.; Gupta, V.; Kallus, N. Data-driven robust optimization. *Math. Program.* **2018**, *167*, 235–292. [CrossRef]
15. Tan, X.; Gong, Z.; Chiclana, F.; Zhang, N. Consensus modeling with cost chance constraint under uncertainty opinions. *Appl. Soft Comput.* **2018**, *67*, 721–727. [CrossRef]
16. Charnes, A.; Cooper, W.W. Chance-constrained programming. *Manag. Sci.* **1959**, *6*, 73–79. [CrossRef]
17. Pagnoncelli, B.; Ahmed, S.; Shapiro, A. Sample average approximation method for chance constrained programming: theory and applications. *J. Optim. Theory Appl.* **2009**, *142*, 399–416. [CrossRef]
18. Kim, J.; Do Chung, B.; Kang, Y.; Jeong, B. Robust optimization model for closed-loop supply chain planning under reverse logistics flow and demand uncertainty. *J. Clean. Prod.* **2018**, *196*, 1314–1328. [CrossRef]
19. Lotfi, S.; Zenios, S.A. Robust VaR and CVaR optimization under joint ambiguity in distributions, means, and covariances. *Eur. J. Oper. Res.* **2018**, *269*, 556–576. [CrossRef]
20. Goh, J.; Sim, M. Distributionally robust optimization and its tractable approximations. *Oper. Res.* **2010**, *58*, 902–917. [CrossRef]
21. Zhu, Z.; Zhang, J.; Ye, Y. Newsvendor optimization with limited distribution information. *Optim. Methods Softw.* **2013**, *28*, 640–667. [CrossRef]
22. Cheung, S.S.; Man-Cho So, A.; Wang, K. Linear matrix inequalities with stochastically dependent perturbations and applications to chance-constrained semidefinite optimization. *SIAM J. Optim.* **2012**, *22*, 1394–1430. [CrossRef]
23. Zymler, S.; Kuhn, D.; Rustem, B. Distributionally robust joint chance constraints with second-order moment information. *Math. Program.* **2013**, *137*, 167–198. [CrossRef]
24. Marshall, A.W.; Olkin, I. A one-sided inequality of the Chebyshev type. *Ann. Math. Stat.* **1960**, *31*, 488–491. [CrossRef]
25. Godwin, H. On generalizations of Tchebychef's inequality. *J. Am. Stat. Assoc.* **1955**, *50*, 923–945. [CrossRef]
26. Mora, N. What determines creditor recovery rates? *Econ. Rev.* **2012**, *97*, 79–109. Available online: https://econpapers.repec.org/article/fipfedker/y_3a2012_3ai_3aqii_3an_3av.97no.2_3ax_3a2.htm (accessed on 29 June 2019).
27. Khieu, H.D.; Mullineaux, D.J.; Yi, H.C. The determinants of bank loan recovery rates. *J. Bank Financ.* **2012**, *36*, 923–933. [CrossRef]
28. Acharya, V.V.; Bharath, S.T.; Srinivasan, A. Does industry-wide distress affect defaulted firms? Evidence from creditor recoveries. *J. Financ. Econ.* **2007**, *85*, 787–821. [CrossRef]
29. Emery, K.; Cantor, R.; Arner, R. Recovery Rates on North American Syndicated Bank Loans, 1989–2003. Available online: https://www.moodys.com/sites/products/DefaultResearch/2006600000428092.pdf (accessed on 30 April 2019).
30. Asarnow, E.; Edwards, D. Measuring loss on defaulted bank loans: A 24-year study. *J. Com. Lend.* **1995**, *77*, 11–23.
31. Gupton, G.M.; Finger, C.C.; Bhatia, M. *CreditMetrics: Technical document*; JP Morgan: New York, NY, USA, 1997. Available online: http://homepages.rpi.edu/~guptaa/MGMT4370.09/Data/CreditMetricsIntro.pdf (accessed on 29 June 2019).
32. Morgan, J. *Creditmetrics-Technical Document*; JP Morgan: New York, NY, USA, 1997. Available online: https://www.msci.com/documents/10199/93396227-d449-4229-9143-24a94dab122f (accessed on 28 April 2019).
33. Rockafellar, R.T.; Uryasev, S. Optimization of conditional value-at-risk. *J. Risk* **2000**, *2*, 21–42. [CrossRef]
34. Sarkar, B. Mathematical and analytical approach for the management of defective items in a multi-stage production system. *J. Clean. Prod.* **2019**, *218*, 896–919. [CrossRef]

© 2019 by the authors. Licensee MDPI, Basel, Switzerland. This article is an open access article distributed under the terms and conditions of the Creative Commons Attribution (CC BY) license (http://creativecommons.org/licenses/by/4.0/).

Article

Customer Exposure to Sellers, Probabilistic Optimization and Profit Research

Miltiadis Chalikias [1], Panagiota Lalou [1] and Michalis Skordoulis [2],*

[1] Department of Accounting and Finance, School of Business and Economics, University of West Attica, 250 Thivon and Petrou Ralli, GR12244 Egaleo, Greece
[2] Department of Forestry and Management of the Environment and Natural Resources, School of Agricultural and Forestry Sciences, Democritus University of Thrace, 193 Pantazidou, GR68200 Orestiada, Greece
* Correspondence: mskordou@fmenr.duth.gr

Received: 6 June 2019; Accepted: 10 July 2019; Published: 12 July 2019

Abstract: This paper deals with a probabilistic problem in which there is a specific probability for a customer to meet a seller in a specified area. It is assumed that the area in which a seller acts follows an exponential distribution and affects the probability of meeting with a customer. Furthermore, the range in which a customer can meet a seller is another parameter which affects the probability of a successful meeting. The solution to the problem is based on a bomb fragmentation model using Lagrange equations. More specifically, using Lagrange equations, the abovementioned dimensions will be calculated in order to optimize the probability of a customer meeting a seller.

Keywords: probabilistic analysis; optimization; Lagrange equations; operations research; bomb fragmentation

1. Introduction

Probabilistic optimization methods may have several applications in many scientific areas, such as business management, finance, computer science, nuclear safety, and the environment [1]. Thus, probabilistic optimization methods have been used by several researchers for various cases. In the field of business, probabilistic optimization models can be used in cases such as service identification, process standardization [2], and sales modeling [3]. These methods can be useful in many decision-making cases in business as they are built upon uncertainty [4].

Based on a primary analysis of the same context [5], the aim of this study is to further analyze the range in which a seller acts as well as the range in which customers can meet with a seller and buy the products they sell. The aim of this analysis is the probability optimization of a successful meeting between a seller and a customer. The warfare problems introduced by Finn and Kent [6] and Przemieniecki [7] will be the basis of the analysis due to the fact that such problems are applied in many business cases [8,9].

The model to be developed is based on Lagrange multipliers and concerns the optimization of a cluster bomb's probability to destroy various enemy targets. The main assumptions of the mathematical model is that the area to be hit by the cluster bomb is a circle with radius R and the bomb's clusters are normally distributed [10].

2. Probabilistic Model Formulation

Based on the abovementioned model, we assume that a number of sellers (N) are spread in a circular area with radius R (e.g., the center of a park). X_1 and X_2 are assumed to be the positions of the sellers in the examined area. The aim is to calculate the maximum probability of a customer being in the area to be met with a seller.

The dispersion of each seller in the area can be a normally distributed two-dimensional random variable following the probability:

$$f(x_1, x_2) = \frac{1}{2\pi\sigma_1\sigma_2\sqrt{1-p^2}} e^{-\varphi(x_1, x_2)}$$

where p refers to the Pearson correlation coefficient and $\varphi(x_1, x_2)$ is

$$\varphi(x_1, x_2) = \frac{1}{2(1-p^2)}\left[\left(\frac{x_1-\mu_1}{\sigma_1}\right)^2 - 2p\left(\frac{x_1-\mu_1}{\sigma_1}\right)\left(\frac{x_2-\mu_2}{\sigma_2}\right) + \left(\frac{x_2-\mu_2}{\sigma_2}\right)^2\right]$$

Assuming that $p = 0$, $\mu_1 = \mu_2 = 0$ and, $\sigma_1 = \sigma_2 = \sigma$, we obtain

$$f(x_1, x_2) = \frac{1}{2\pi\sigma^2} e^{-\frac{x_1^2 + x_2^2}{2\sigma^2}}$$

We will now calculate the probability of a customer being inside the abovementioned circular area. Thus, the following transformations are made:

$$x_1 = r\cos\theta$$
$$x_2 = r\sin\theta$$

where

$$0 \leq r \leq R$$
$$0 \leq \theta \leq 2\pi$$

and

$$\det\left(\frac{\partial(x_1, x_2)}{\partial(r, \theta)}\right) = \begin{vmatrix} \cos\theta & -r\sin\theta \\ \sin\theta & r\cos\theta \end{vmatrix} = r$$

Therefore,

$$\iint_D f(x_1, x_2)dx_1 dx_2 = \int_0^{2\pi}\int_0^R \frac{1}{2\pi\sigma^2} e^{-\frac{r^2}{2\sigma^2}} r\, dr d\theta = 1 - e^{-\frac{R^2}{2\sigma^2}}$$

Thus, the probability of a potential customer being in the seller's area is

$$P_1 = 1 - e^{-\frac{R^2}{2\sigma^2}}$$

where R refers to the seller's action area radius.

We assume a radius (R) of a circular area in which only half of the sellers are located. The value of R where $P_1 = 0,5$ is denoted by (c). Then, we obtain

$$P_1 = 0,5$$

$$\Leftrightarrow 1 - e^{-\frac{c^2}{2\sigma^2}} = 0,5$$

$$\Leftrightarrow e^{-\frac{c^2}{2\sigma^2}} = \frac{1}{2}$$

$$\Leftrightarrow \frac{c^2}{2\sigma^2} = \ln 2$$

$$\Leftrightarrow 2\sigma^2 = \frac{c^2}{\ln 2}$$

Now, we obtain
$$P_1 = 1 - e^{-\frac{R^2 \ln 2}{c^2}}$$

We now consider A to be the area in which a customer is exposed to a seller. The probability that the customer is not in this area is
$$1 - \frac{A}{\pi R^2}$$

Based on the first assumption of the analysis that the seller number is equal to N, the probability that a customer does not meet with any of the sellers is
$$\left(1 - \frac{A}{\pi R^2}\right)^N$$

We can state
$$\left(1 - \frac{A}{\pi R^2}\right)^N \approx e^{-\frac{AN}{\pi R^2}}$$

Thus, the probability that a customer meets with at least one of the sellers is
$$P = \left(1 - e^{-\frac{R^2 \ln 2}{c^2}}\right)\left(1 - e^{-\frac{AN}{\pi R^2}}\right)$$

In order to maximize this probability, we consider
$$x = \frac{R^2 \ln 2}{c} \text{ and, } y = \frac{AN}{\pi R^2} \qquad (1)$$

Therefore, the probability function is
$$P(x, y) = (1 - e^{-x})(1 - e^{-y})$$

We note that
$$xy = \frac{R^2}{c^2} \ln 2 \frac{AN}{\pi R^2} = \frac{\ln 2 AN}{c^2} = \text{constant}$$

We now assume that $xy = k$. Thus, we must calculate the extreme values of the function:
$$P(x, y) = (1 - e^{-x})(1 - e^{-y})$$

under the restriction $g(x, y) = xy - k = 0$.

We will use the Lagrange multipliers method. We define the Lagrange function as follows:
$$L(x, y, \lambda) = P(x, y) + \lambda g(x, y)$$
$$\Leftrightarrow L(x, y, \lambda) = (1 - e^{-x})(1 - e^{-y}) + \lambda xy - \lambda k$$

In order to find the critical points, we solve the system of the following equations:
$$\begin{cases} \frac{\partial L}{\partial x} = 0 \Leftrightarrow e^{-x}(1 - e^{-y}) = -\lambda y \\ \frac{\partial L}{\partial y} = 0 \Leftrightarrow e^{-y}(1 - e^{-x}) = -\lambda x \\ \frac{\partial L}{\partial \lambda} = 0 \Leftrightarrow xy = k \end{cases}$$

which deduces to:
$$x = y$$

Afterwards, we calculate the following determinant:

$$D = \begin{vmatrix} 0 & \frac{\partial^2 L}{\partial \lambda \partial x} & \frac{\partial^2 L}{\partial \lambda \partial y} \\ \frac{\partial^2 L}{\partial x \partial \lambda} & \frac{\partial^2 L}{\partial x^2} & \frac{\partial^2 L}{\partial x \partial y} \\ \frac{\partial^2 L}{\partial y \partial \lambda} & \frac{\partial^2 L}{\partial y \partial x} & \frac{\partial^2 L}{\partial y^2} \end{vmatrix} = \begin{vmatrix} 0 & y & x \\ y & -e^{-x}(1-e^{-y}) & e^{-x}e^{-y}+\lambda \\ x & e^{-x}e^{-y}+\lambda & -e^{-y}(1-e^{-x}) \end{vmatrix}$$

For $x = y$, we obtain

$$D = \begin{vmatrix} 0 & x & x \\ x & -e^{-x}(1-e^{-x}) & e^{-2x}+\lambda \\ x & e^{-2y}+\lambda & -e^{-y}(1-e^{-y}) \end{vmatrix} = 2x^2(e^{-y}+\lambda) > 0$$

The above equations mean that the maximum value of the function $P(x,y)$ at the point (x,y) is when $x = y$. Based on Equation (1), we now obtain

$$\frac{R^2 \ln 2}{c^2} = \frac{AN}{\pi R^2} \Leftrightarrow R = \sqrt[4]{\frac{c^2 AN}{\ln 2\pi}}$$

Thus, the maximum probability of a customer meeting a seller is

$$P_{max} = \left(1 - e^{-\sqrt{\frac{\ln 2 AN}{c^2 \pi}}}\right)^2$$

We now assume that the customer number is equal to k times the number of the sellers. Thus, the mean number of customers to meet sellers is

$$\hat{p} = vP_{max} = kNP_{max}$$

Furthermore, we assume that the mean number of customer and seller meetings follows the Poisson distribution, where

$$\lambda = N\hat{p} = N^2 k P_{max}$$

Thus, the probability for at least one of the customers meeting a seller is

$$1 - P(0) = 1 - e^{-\lambda} = 1 - e^{-kN^2\left(1 - e^{-\sqrt{\frac{\ln 2 AN}{c^2 \pi}}}\right)^2}$$

3. Operational Research Application

We assume that 10% of customer and seller meetings lead to a successful sale with a gain of 0.2 for the seller. Moreover, we assume that the seller has accommodation costs. Therefore, we conclude that the seller's profit is positive when the number of meetings is over 5N.

We can calculate the probability of x meetings taking place by using the formula of Poisson distribution:

$$\lambda = N^2 k\left(1 - e^{-\sqrt{\frac{\ln 2 AN}{c^2 \pi}}}\right)^2$$

In Figure 1, we see that the probability of the number of meetings is x, assuming that the surface area around the seller is $A = 100$, the radius of the circular disc is $c = 100$ and the number of sellers is $N = 10$.

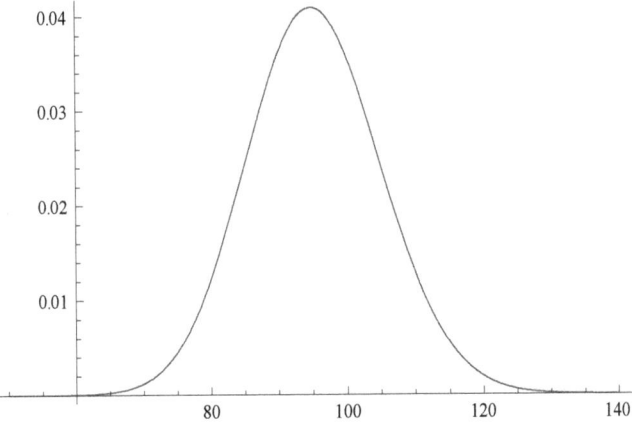

Figure 1. Probability of the number of meetings.

In Tables 1–5, we calculate the probability of customers and sellers meetings to be >5N, so that sellers have profit. We assume that $A = 50m^2$ and $c = 100m$.

Table 1. Number of customers = 10N (k = 10).

N	5N	P(X>5N)
16	80	$1.14847 \cdot 10^{-8}$
17	85	$5.2996 \cdot 10^{-7}$
18	90	0.0000174362
19	95	0.00038038
20	100	0.00516161
21	105	0.0413728
22	110	0.189822
23	115	0.501292
24	120	0.818503
25	125	1

According to the above table, when N is 23, it is more possible for a seller to make a profit (the probability exceeds 50%).

Table 2. Number of customers = 20N (k = 20).

N	5N	P(X>5N)
10	50	$1.03905 \cdot 10^{-8}$
11	55	$7.18658 \cdot 10^{-7}$
12	60	0.0000357685
13	65	0.00108754
14	70	0.0175679
15	75	0.135727
16	80	0.481138
17	85	0.852159
18	90	1

Following the same analysis, we create tables for k = 30, 40 and 50.

Table 3. Number of customers = 30N (k = 30).

N	5N	P(X>5N)
9	45	$9.64421 \cdot 10^{-6}$
10	50	0.000457264
11	55	0.0112619
12	60	0.119434
13	65	0.495747
14	70	0.887378
15	75	1

Table 4. Number of customers = 40N (k = 40).

N	5N	P(X>5N)
9	45	0.00383084
10	50	0.0652488
11	55	0.391667
12	60	0.849736
13	65	1

Table 5. Number of customers = 50N (k = 50).

N	5N	P(X>5N)
9	45	0.093595
10	50	0.502147
11	55	0.921185
12	60	1

4. Conclusions and Future Research

The aim of the above analysis was to propose a mathematical model which calculates the optimal range in which sellers should act in an area as well as the optimal range in which customers can meet with the sellers and buy their products.

The optimization of the probability of customers meeting sellers can be applied in several operational research problems. Due to the fact that, in this paper, a mathematical model is proposed, future research may concern the model's empirical application (e.g., the success of moving advertisements) which would lead to good fit confirmation as well.

Moreover, future research may concern further generalizations of the model assuming different distribution functions than Poisson distribution. More specifically, despite the fact that Poisson or two-dimensional normal distribution are the most relevant distributions in a case such as the one examined [11,12], they are not the only ones. For example, a probabilistic model, such as that used by Bass [13], could be used.

Finally, despite the fact that probabilistic models are built upon uncertainty, which is taken into account, there could be differences between the theoretical models and the observed values due to a variety of potential constraints mainly associated with the external environment [14]. The more variability a probabilistic optimization includes, the better results would be obtained [14]. Thus, the highest possible number of variables that can affect the model should be taken into consideration.

Author Contributions: M.C., P.L. and M.S. equally contributed to the paper's conceptualization. M.C. developed and designed the research methodology and models. M.C. and P.L. performed the experiments, calculations and computer programming. M.S. provided the study materials, verified the reproducibility of the research results and created the published work.

Funding: This research received no external funding.

Conflicts of Interest: The authors declare no conflicts of interest.

References

1. Uryasev, S. *Probabilistic Constrained Optimization: Methodology and Applications*; Springer Science and Business Media: Dordrecht, The Netherlands, 2000; Volume 49.
2. Leopold, H.; Niepert, M.; Weidlich, M.; Mendling, J.; Dijkman, R.; Stuckenschmidt, H. Probabilistic optimization of semantic process model matching. In Proceedings of the Business Process Management 10th International Conference (BPM 2012), Tallinn, Estonia, 3–6 September 2012; Barros, A., Gal, A., Kindler, E., Eds.; Springer Nature: Berlin, Germany, 2012; pp. 319–334.
3. Duncan, B.A.; Elkan, C.P. Probabilistic modeling of a sales funnel to prioritize leads. In Proceedings of the 21th ACM SIGKDD International Conference on Knowledge Discovery and Data Mining, Sydney, Australia, 10–13 August 2015; Association for Computing Machinery: New York, NY, USA, 2015; pp. 1751–1758.
4. Bertsimas, D.; Thiele, A. Robust and data-driven optimization: Modern decision-making under uncertainty. *INFORMS Tutor. Oper. Res.* **2006**, *3*, 95–122.
5. Lalou, P.; Chalikias, M.; Skordoulis, M.; Papadopoulos, P.; Fatouros, S. A probabilistic evaluation of sales expansion. In Proceedings of the 5th International Symposium and 27th National Conference on Operation Research, Piraeus, Greece, 9–11 June 2016; Piraeus University of Applied Sciences: Piraeus, Greece, 2016; pp. 109–112.
6. Finn, M.V.; Kent, G.A. *Simple Analytic Solutions to Complex Military Problems*; Rand Corporation: Santa Monica, CA, USA, 1985.
7. Przemieniecki, J. Mathematical Methods in Defense Analyses. In *Mathematical Methods in Defense Analyses*; American Institute of Aeronautics and Astronautics (AIAA): Reston, VA, USA, 2000.
8. Chalikias, M.; Skordoulis, M. Implementation of Richardson's Arms Race Model in advertising expenditure of two competitive firms. *Appl. Math. Sci.* **2014**, *8*, 4013–4023. [CrossRef]
9. Chalikias, M.; Lalou, P.; Skordoulis, M. Modeling Advertising expenditures using differential equations: The case of an oligopoly data set. *J. Appl. Math. Stat.* **2016**, *55*, 14–22.
10. Daras, N.I. *Operational Research and Military Applications: Strategic Defense*; Hellenic Arms Control Center: Athens, Greece, 2007.
11. Jin, T.; Taboada, H.; Espiritu, J.; Liao, H. Allocation of reliability–redundancy and spares inventory under Poisson fleet expansion. *IISE Trans.* **2017**, *49*, 737–751. [CrossRef]
12. Gaur, V.; Kesavan, S. The Effects of Firm Size and Sales Growth Rate on Inventory Turnover Performance in the U.S. Retail Sector. In *Handbook of Global Logistics*; Springer Science and Business Media LLC: Berlin, Germany, 2015; Volume 223, pp. 25–52.
13. Bass, F.M. A new product growth for customer durables. *Manag. Sci.* **1969**, *15*, 215–227. [CrossRef]
14. Yoshimura, J.; Shields, W.M. Probabilistic optimization of phenotype distributions: A general solution for the effects of uncertainty on natural selection? *Evol. Ecol.* **1987**, *1*, 125–138. [CrossRef]

© 2019 by the authors. Licensee MDPI, Basel, Switzerland. This article is an open access article distributed under the terms and conditions of the Creative Commons Attribution (CC BY) license (http://creativecommons.org/licenses/by/4.0/).

Article

Digital Supply Chain through Dynamic Inventory and Smart Contracts

Pietro De Giovanni

Department of Business and Management, LUISS University, 00197 Rome, Italy; pdegiovanni@luiss.it;
Tel.: +39-0685225935

Received: 9 November 2019; Accepted: 7 December 2019; Published: 13 December 2019

Abstract: This paper develops a digital supply chain game, modeling marketing and operation interactions between members. The main novelty of the paper concerns a comparison between static and dynamic solutions of the supply chain game achieved when moving from traditional to digital platforms. Therefore, this study proposes centralized and decentralized versions of the game, comparing their solutions under static and dynamic settings. Moreover, it investigates the decentralized supply chain by evaluating two smart contracts: Revenue sharing and wholesale price contracts. In both cases, the firms use an artificial intelligence system to determine the optimal contract parameters. Numerical and qualitative analyses are used for comparing configurations (centralized, decentralized), settings (static, dynamic), and contract schemes (revenue sharing contract, wholesale price contract). The findings identify the conditions under which smart revenue sharing mechanisms are worth applying.

Keywords: digital supply chain; smart contracts; dynamic inventory; revenue sharing contract

1. Introduction

Despite its recent advent, supply chain (SC) management was deeply investigated in terms of vertical and horizontal relationships, multitude of tactics and strategies, and various objectives, purposes, and targets, mainly based on the maximization of profit or minimization of costs [1]. Although several contributions emphasized the need for implementing SC strategies, research is still needed to identify the best SC structure and the appropriate methodology for its analysis, especially with new technological disruptions like digitalization, Industry 4.0, and blockchains, which are deeply changing operations and supply chain management. From a methodological point of view, the literature applies both static and dynamic methods to investigate the supply chain management relationships.

The static literature involves time-independent features. In each period, the models do not observe the changes in system parameters, and the main characterization involves decisions on customer demand (deterministic and random, endogenous and exogenous), vertical and horizontal competition within supply chains, and no risk involved, risk incurred by only one or few members, or risk shared between the participants [2].

According to the static literature, the general concept of Nash equilibrium applies for determining the optimal solution, allowing firms to maximize a certain payoff. Starting from this general characterization, the literature proposes multitude types of scenarios including cooperative and non-cooperative settings, simultaneous and sequential decision-making (Cournot and Stackelberg games), vertical and horizontal competition, and embracing numerous areas as pricing (Bertrand's competition) or production (Cournot's competition).

In contrast, the dynamic literature considers how the supply chain relationships change and evolve over time. Several subjects can be dynamically analyzed, like the word-of-mouth effect, economies of scale, uncertainty, the bull whip effect, and seasonal, fashion, and holiday demand pattern. In this framework, the system changes can occur at any point of time; consequently, the control of actions and the monitoring of strategies need continuous investigation [3]. SCs adjust their decisions over time, resulting in an inter-temporal interaction among supply chain members. Dynamic setting incorporates a carry-over effect. For example, advertising not only affects the current sales, but it also feeds the brand equity, which in turn has an influence on sales [4]. Ref. [5] proposed the application of a dynamic approach to solve the dilemma between short- and long-term coordination. The first application of a differential game was developed by [6]; afterward, a number of contributions surveyed applications in several areas [7–18].

As highlighted by [17], although analyzing supply chains in a static framework generates myopic attitudes, in reality, it remains an ideal solution for supply chain coordination under particular conditions. Similarly, some static mechanisms also coordinate the SC in dynamic settings, whereas others do not [19]. This research contributes the described framework in which the effectiveness of static and dynamic approaches for supply chain coordination is not clear and the boundaries of applications are not clearly defined. According to the literature, several coordination schemes exist. Each of them possesses particular characteristics, and their adoption depends on both the SC structures and the inter-temporal setting. Previous literature showed numerous contributions identifying various contract schemes for supply chain coordination as sales rebates [20], buy-back [21], quantity flexibility [22], quantity premium contract [23], wholesale price and revenue sharing contract [21], and quality targets [16].

To provide both theoretical and practical evidence, this research compares the results of two smart contracts, that is, wholesale price (WPC) and revenue sharing contracts (RSC) in both static and dynamic settings. Accordingly, we are able to highlight both the advantages and the disadvantages of each approach to achieve supply chain coordination [24]. Only few contributions studied the differences between static and dynamic frameworks, as well as their limits and applications [2,25]. Moreover, according to [21], wholesale price and revenue sharing contracts generate different and controversial results depending on the SC structure. By setting a static framework, the authors showed that a revenue sharing coordinates a supply chain including one supplier and one retailer, while it fails in the case of multiple retailers and, consequently, WPC is preferable. Each player prefers RSC with respect to WPC whenever the revenues shared exceed the revenues lost due to a lower transferring price. Nevertheless, RSC is difficult to administer and extremely costly; the contract designer could prefer the implementation of a simple and traditional contract coordinating the SC even when it generates some negative externalities.

This research extends the study by [21] by proposing a dynamic version of their model and including the digitalization of the supply chain [24]. The findings take position in the literature by comparing RSC and WPC inside static and dynamic frameworks. Figure 1 reports the adopted SC structure which appeared in [6] and, more recently, in [19].

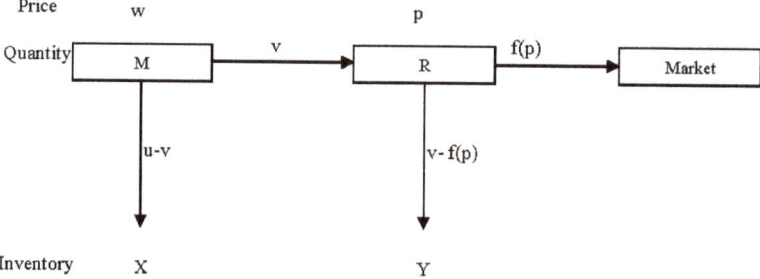

Figure 1. Representation of the supply chain structure.

The SC structure consists of one manufacturer (firm *he*) and one retailer (firm *she*). The manufacturer makes the quantity u and sells the quantity v to the retailer at a price w. The difference between these two quantities represents his inventory X. The retailer purchases v and sells $f(p)$ to the market. $f(p) = a - bp$ represents the inverse demand function adopted by [6,26], where a and b are positive constants. The difference between v and $f(p)$ is the retailer's inventory Y.

The retailer fixes the price while the wholesale price is exogenous and intended as determined by an artificial intelligence system that supports the decision-making process. Under RSC, both players share some revenue through the parameter φ and determine the transferring price w_0 such that $w_0 < w$. Firms can use artificial intelligence to identify the best parameters for maximizing their profit functions [27], complementing the system with other technologies like blockchain and big data. Therefore, the contracts become smart contracts.

Beyond presenting and comparing several decentralized SCs characterized by the adoption of WPC and RSC under static and dynamic frameworks, this research proposes the centralized solution, which is one main element of a digital supply chain. Industry 4.0 technologies allow firms to be fully integrated, and the centralized solution is the best formulation representing this framework. In fact, the artificial intelligence identifies the best contract terms according to the best possible outcomes of a supply chain, which is clearly given by the centralized solution. Hence, the centralized SC is a benchmark [27].

This research is organized as follows: Section 2 describes the differences between static and dynamic games. Section 3 introduces the static formulation of the game proposing both centralized and decentralized solutions. Section 4 develops the dynamic version of the theoretical game following the same steps of Section 3, i.e., centralized and decentralized solutions first and their qualitative comparison afterward. Section 5 proposes numerical analysis of the models developed in Sections 2 and 3, while the last section concludes and presents future directions.

2. Technical Differences between Static and Dynamic Games

The existing literature on inter-firm relationships reports extensively on a large number of theoretical games using both static and dynamic modeling. Essentially, the choice of their adoption depends on the nature of the game. On one hand, static models are time-independent. Any change in the parameters of the system is completely ignored, thereby generating myopic solutions [25]. This class of game is mainly used to characterize decisions on customer demand (deterministic and random, endogenous and exogenous), vertical and horizontal competition, and no-risk actions [21]. On the other hand, a dynamic setting appears to be considerably more appropriate when modeling variables characterized by carry-over effects. For instance, advertising not only affects current sales considerably, but it also feeds the brand equity, which in turn has an influence on sales [28]. Other diffused examples are represented by the "word-of-mouth" effect, economies of scale, uncertainty, the "bull whip" effect, and seasonal, fashion, and holiday demand patterns [21]. In these frameworks, changes may occur at any point in time; consequently, the control of actions is investigated continuously [15].

A static game involves players choosing several strategies simultaneously (one-shot game) or sequentially (strategies are chosen when needed). The static game is cooperative as the players make side-payments or form coalitions [21]. Nash formulates the solution for a non-cooperative game, in which each player maximizes his own objective or payoff function.

A pair of strategies u_1 and u_2 is a Nash equilibrium (NE) if, for every $u_1 \in U_1$ and $u_2 \in U_2$, the following pair of inequalities holds:

$$J^1(u_1^*, u_2^*) \geq J^1(u_1, u_2^*), \text{ and } J^2(u_1^*, u_2^*) \geq J^2(u_1^*, u_2).$$

No players can do better or get more by choosing an action different from ui*, and no player has incentives for choosing any other strategy. In particular, the Nash solution becomes

$$y_1^* = \underset{y_1 \in Y_1}{\operatorname{argmax}}\{J^1(u_1, u_2^*)\} \text{ and } y_2^* = \underset{y_2 \in Y_2}{\operatorname{argmax}}\{J^2(u_1^*, u_2)\}.$$

If the problem is static, strategy sets are not constrained, and the payoff functions are continuously differentiable [21]. First-order conditions apply for finding the best response functions. The NE needs the resolution of a system of best responses translated into a system of first order (necessary) in optimal conditions.

$$\left.\frac{\partial J^1(u_1, u_2^*)}{\partial u_1}\right|_{u_1=u_1^*} = 0 \text{ and } \left.\frac{\partial J^2(u_1^*, u_2)}{\partial u_2}\right|_{u_2=u_2^*} = 0.$$

Moreover, the payoffs are maximized when the second-order condition of (sufficient) optimality holds:

$$\left.\frac{\partial J^1(u_1, u_2^*)}{\partial u_1}\right|_{u_1=u_1^*} < 0 \text{ and } \left.\frac{\partial J^2(u_1^*, u_2)}{\partial u_2}\right|_{u_2=u_2^*} < 0.$$

The NE uses the concept of best response for finding the solution of the game and for maximizing the players' payoffs. Each player chooses the best available action, which generally depends on the other players' actions. In this sense, whenever a player chooses his own actions, he should have in mind the possible actions that the other players might choose (belief). Consequently, the Nash solution changes to

$$y_1^R(y_2) = \underset{y_1 \in Y_1}{\operatorname{argmax}}\{J^1(u_1, u_2)\} \text{ and } y_2^R(y_1) = \underset{y_2 \in Y_2}{\operatorname{argmax}}\{J^2(u_1, u_2)\}.$$

Optimal conditions change accordingly. For further information on the static approach, see the studies by [21,29].

Similarly, dynamic games or differential games involve at least two players having different objective functions. The existing models embrace non-cooperative and cooperative games; however, in both cases, players' interrelationships and coordination change over time. Both players select the values of their control and state variables, and the game varies according to one or more differential equations. In the case of a two-player game ($i = 1, 2$), player 1 chooses his control variable, u_1, in order to maximize

$$J^1(u_1, u_2) = \int_{t_0}^{t_1} f^1[t, x_1(t), x_2(t), u_1(t), u_2(t)]dt,$$

while player 2 maximizes

$$J^2(u_1, u_2) = \int_{t_0}^{t_1} f^2[t, x_1(t), x_2(t), u_1(t), u_2(t)]dt,$$

with both maximization problems subject to the following state equations:

$$\dot{x}_i = g^i[t, x_1(t), x_2(t), u_1(t), u_2(t)]$$
$$x_i(t_0) = x_{i0}, \quad x_i(t_i) \text{ free}.$$

The standard problem assumes the controls to be continuous piecewise, while g^i and f^i are both known and continuous differentiable functions in the four arguments.

Starting from this general formulation, various researches developed the concept of NE in a dynamic setting, occurring whenever

$$J^1(u_1^*, u_2^*) > J^1(u_1, u_2^*), \text{ and } J^2(u_1^*, u_2^*) > J^2(u_1^*, u_2),$$

where u_1^* and u_2^* describe the optimal control variables of the players. Two mainstream strategies are applied in differential games: Open and closed (feedback) strategies. In the open-loop strategies, each player selects the values of his control variable at the outset of the game, the instantaneous value of which is only a function of time. Players commit their actions at the beginning of the game without making any modification; they do not update the game according to the available information [15].

Hamiltonian and necessary conditions for optimality easily allow the determination of the NE open-loop strategy. For $i = 1, 2$ and $j = 1, 2$, the standard optimal control analysis involves

$$H^i\left[t, x_1(t), x_2(t), u_1(t), u_2(t), \lambda_1^i(t), \lambda_2^i(t)\right]$$
$$H_{u_i}^i = 0$$
$$\dot{\lambda}_j^i = -\partial H^i / \partial x_i$$

In reality, players commit to the entire course of actions over time, revising their actions as the game instantaneously evolves, according to the state of the system. This class of strategies coincides with the closed-loop strategies.

The instantaneous control is a function not only of time but also of the state of the system $x_1(t)$, $x_2(t)$ at that time. In this sense, the closed-loop strategy result is a perfect subgame. The game continues at a new stage representing a subgame of the original one. The state of the system variables evolves accordingly to the new state. A feedback strategy allows the players to take their best decisions even if the initial state of the subgame evolves under suboptimal actions, revealing optimality for the initial game, as well as for every subgame evolving from it.

Previous studies did not apply that class of strategy as frequently as the open-loop strategy. Numerous computational difficulties exist. Even when applying Hamiltonian and first-order conditions, the closed-loop games require the computation of the other players' feedback strategy. For $i = 1, 2$ and $j = 1, 2$, the feedback strategy requires the following conditions:

$$H^i\left[t, x_1(t), x_2(t), u_1(t, x_1(t), x_2(t)), u_2(t, x_1(t), x_2(t)), \lambda_1^i(t), \lambda_2^i(t)\right]$$
$$H_{u_i}^i = 0$$
$$\dot{\lambda}_j^1 = -\partial H^1 / \partial x_j - (\partial H^1 / \partial u_2)(\partial u_2^* / \partial x_j)$$
$$\dot{\lambda}_j^1 = -\partial H^2 / \partial x_j - (\partial H^2 / \partial u_1)(\partial u_1^* / \partial x_j)$$

The second part of the right-hand side of the co-states identifies the interaction term. It also requires the computation of partial differential equations; hence, the optimal solution is not guaranteed. Finding player 1's optimal feedback strategy requires knowing players 2's optimal feedback strategy, which requires knowing player 1's feedback strategy, and so on. However, the closed-loop strategy describes the real competition and the firms' interactions more appropriately. For further information on the dynamic approach, see the studies by [21].

When players adopt the optimal open-loop concept, they design the optimal strategy at the beginning of the planning horizon and then stick to it until the end. By contrast, the term "memoryless" characterizes closed-loop strategies that change over time as new information becomes available [25].

Table 1 summarizes the main differences between static and dynamic games. Static games mainly apply for single (one-shot) games. Their application to the supply chain (SC) does not seem appropriate since long-term and cooperative relationships characterize those games. Nevertheless, the literature shows numerous studies applying static games to short-term and single-period relationships and reporting myopic solutions. In static games, the optimal value of the players' decision variables is derived by developing a system of first-order conditions. Time does not influence any of them. Static games use the concept of best response for finding the most favorable outcomes for the players, taking other players' strategies as given [21]. They search for the NE, that is, the point at which each player selects his best response with respect to the other players' strategies in order to maximize his payoff.

Table 1. Technical differences between static and dynamic games.

	Static	Dynamic
Planning horizon	One-shot game (t = 1)	The games have starting (t_0) and ending (t) periods, or are even infinite
Types of games	Cooperative and non-cooperative	Cooperative and non-cooperative over open- and closed-loop frameworks
Mathematical computations required	System of first-order conditions	System of ordinary (partial) differential equations in open (closed)-loop games
Equilibrium	Best response function	Optimal control variable
General objective	Player payoff maximization according to the best response function	Player payoff maximization according to the optimal control variable and the optimal trajectory of the state variables

Dynamic modeling fits adequately when modeling long-term inter-firm games. By choosing the optimal control variables, dynamic modeling describes the optimal trajectory of the state variables that optimizes the players' objective functions. Differential games adequately model cooperative and non-cooperative settings. Each rule is defined from the beginning of the game, and players keep or modify it according to the adoption of an open- or a closed-loop strategy. An open-loop strategy, in fact, definitely fits when modeling integrated SCs that are characterized by diffused trust, collaboration, and commitment. A closed-loop strategy, in contrast, models SCs that are not at all integrated, as well as partnerships, alliances, and other forms of inter-firm relationships, in which the partners develop long-term strategies that are mainly driven by coordination [19].

Static and dynamic methods are extensively applied to firm games, modeling cooperative and non-cooperative scenarios. They cover several areas and show theoretical and practical results. The next section introduces the state of the art in static and dynamic games with particular attention to SC games. This class of game, in fact, received significant attention recently, while monopolist and oligopolistic scenarios rarely appear in today's world of business.

3. Characterization of the Static Game

In the formulation of the static model, the manufacturer and retailer maximize their own profit functions, π^m and π^r, respectively. We refer to this model as an SC game, because the two firms compete with each other in the maximization of their profits. Therefore, each firm sets the optimal strategies to maximize its own profit function rather than maximizing the profit function of the entire chain. Equations (1) and (2) illustrate these functions.

$$\pi^m = (1-\varphi)(a-bp)p + wv - c_m \frac{(u-\bar{u})^2}{2} - h_m(\hat{X} - v + u). \qquad (1)$$

$$\pi^r = \varphi(a-bp)p - wv - c_r \frac{(v-\bar{v})^2}{2} - h_r[\hat{Y} + v - f(p)]. \qquad (2)$$

The retailer benefits from total revenue $R(p,v) = (a-bp)p$ that is shared with the manufacturer depending on the contract scheme adopted. According to Equations (1) and (2), the parameter φ describes the sharing of revenues between the two players under RSC that is determined by an artificial intelligence system that supports the firms' contractual development. In this latter case, the sharing parameter shows values comprised in the interval (0, 1). As its value increases, the amount of revenue transferred from the retailer to the manufacturer decreases. In contrast, as its value approaches 0, the revenue shared with the manufacturer increases. $\varphi = 1$ reflects the wholesale price scheme, in which the manufacturer does not obtain any share of the retailer's revenue.

This research adopts similar production and purchasing functions as those introduced by [26]. In particular, the manufacturer's production and the retailer's purchasing cost functions take a quadratic form, $c_m(u-\bar{u})^2/2$ and $c_r(v-\bar{v})^2/2$, respectively. c_m (c_r) represents the marginal production (purchasing) cost, while the total cost has a quadratic form; as the quantity produced (purchased) increases, the total production (purchasing) cost decreases. $\bar{u}(\bar{v})$ represents the optimal production (purchasing) quantity for exploiting economies of scale and minimizing the total production (purchasing) costs. The convexity of the quadratic cost function imposes penalties when u and v diverge from the target level $\bar{u}$ and $\bar{v}$.

The last part of the profit functions concerns the holding inventory costs. h_m and h_r are the manufacturer's and retailer's marginal inventory costs, while expressions $(\hat{X}-v+u)$ and $\left[\hat{Y}+v-f(p)\right]$ represent the quantities held in stock by the manufacturer and the retailer, respectively. The manufacturer's stock includes the initial quantity $\hat{X}$, the production rate u, and the retailer's purchasing rate v. The retailer's stock accounts for the initial quantity $\hat{Y}$, the purchasing rate v, and the quantity sold $f(p)$.

As long as this research characterizes the static game, it presents the existence of the NE as a solution of the previous system. It is assumed that the manufacturer entirely satisfies the quantity demanded by the retailer and, similarly, the retailer entirely satisfies the demand by the market; hence, $u > v > f(p)$. Assuming that the transferring price w is given and exogenous, u denotes the manufacturer's decision variable, while v and p indicate the retailer's one. The NE is a pair of strategies $\sigma^m(u^*,v^*,p^*)$ and $\sigma^r(u^*,v^*,p^*)$, such that neither the manufacturer nor the retailer deviate.

In particular:

$$\sigma^m(u^*,v^*,p^*) > \sigma^m(u,v^*,p^*) \quad \forall v,u,p. \tag{3}$$

$$\sigma^r(u^*,v^*,p^*) > \sigma^r(u^*,v,p) \quad \forall v,u,p. \tag{4}$$

The existence of the Nash equilibrium is checked through verifying the concavity of the player's payoff. Suppose that the strategy space is compact and convex, and the payoff function is continuous and quasi-concave with respect to each player's strategy. There exists at least one Nash equilibrium in pure strategy. Easily, the quasi-concavity of the profit functions applies for both players. Given a parameter φ such that $0 < \varphi < 1$, concavity is generally given by

$$\pi^i[\varphi(u\prime,v\prime,p\prime)+(1-\varphi)(u'',v'',p'')] > \varphi\pi^i(u\prime,v\prime,p\prime)+(1-\varphi)\pi^i(u'',v'',p''). \tag{5}$$

Proposition 1. *Quasi-concavity of the profit function is a necessary condition for optimality.*

Proof. Since the quasi-concavity is expressed by

$$\pi^i[\varphi(u\prime,v\prime,p\prime)+(1-\varphi)(u'',v'',p'')] < \max[\pi^i(u\prime,v\prime,p\prime),\pi^i(u'',v'',p'')], \tag{6}$$

the necessary condition for optimality in Equation (5) holds. For each $0 < \varphi < 1$, the linear combination of one pair of profit functions is always lower than the maximum profit generated by one of the two strategies. □

As proposed by [1], we may use the implicit function theorem for investigating the uniqueness of the equilibrium.

$$\sum_{i=1, i\neq k}^{n} \left|\frac{\partial^2 \pi_i}{\partial u_i \partial u_k}\right| < \left|\frac{\partial^2 \pi_k}{\partial u_k^2}\right| \quad \forall k. \tag{7}$$

Proposition 2. *The equilibrium of the game is unique with respect to the production and purchasing rate.*

Proof. $\frac{\partial^2 \pi_i}{\partial u_i \partial u_k}$ is zero for both players since the variables at the denominator are independent, while the absolute value of $\frac{\partial^2 \pi_k}{\partial u_k^2}$ is higher than zero; therefore, the resulting equilibrium exists and is unique. □

3.1. Centralized Solution of the Static SC Game

When studying the centralized SC, the original objective functions in Equations (1) and (2) sum up in a unique one. The game becomes a Pareto static game [30]. Several contributions introduced the profit function in centralized SC. [2] introduced the centralized SC composed by one supplier and one manufacturer, showing decentralized and centralized profit functions in RSC in a static game. [25] proposed centralized and decentralized profit functions in static and dynamic settings for Stalkeberg and Cournot games. To get the centralized profit function, the firms' profit functions in Equations (1) and (2) are summed up, implying the elimination of the transferring value w_0v. Finally, the centralized SC profit function results.

$$\pi^c = (a-bp)p - c_m \frac{(u-\bar{u})^2}{2} - c_r \frac{(v-\bar{v})^2}{2} - h_m(\hat{X} - v + u) - h_r[\hat{Y} + v - f(p)]. \tag{8}$$

In the centralized formulation of the static model, the firms behave as one player working as a single entity. The first order condition for the decision variables is as follows:

$$\frac{\partial \pi^c}{\partial u} = -c_m(u - \bar{u}) - h_m = 0 \quad \Rightarrow u = \bar{u} - \frac{h_m}{c_m}, \tag{9}$$

$$\frac{\partial \pi^c}{\partial v} = -c_r(v - \bar{v}) - h_r + h_m = 0 \quad \Rightarrow v = \bar{v} + \frac{h_m - h_r}{c_r}, \tag{10}$$

$$\frac{\partial \pi^c}{\partial p} = a - 2bp - h_r b = 0 \quad \Rightarrow p = \frac{1}{2}\left(\frac{a}{b} - h_r\right). \tag{11}$$

Substituting Equations (9)–(11) into Equation (8), the profit function in centralized SC becomes

$$\pi^c = \frac{1}{4}\left(\frac{a^2}{b} - bh_r^2\right) + \frac{h_m^2}{2c_m} + (h_m - h_r)\left(\bar{v} + \frac{h_m - h_r}{2c_r}\right) - h_m(\hat{X} + \bar{u}) - h_r\left(\hat{Y} - \frac{1}{2}(a + h_rb)\right). \tag{12}$$

Equation (12) is the benchmark of the SC game under static settings. Whatever the coordination scheme adopted under decentralized settings, Equation (12) shows always higher profits.

The centralized solution is a positive benchmark since it allows firms to identify the best outcomes that they can get. The artificial intelligence system uses the centralized solution to search for the best contract terms to be used in supply chain coordination.

3.2. Decentralized Solution of the Static SC Game

In order to characterize the equilibrium with respect to the manufacturer's and retailer's decision variables and finding the solution of the decentralized SC, this research develops the necessary conditions for optimality of both Equation (1) and Equation (2).

$$\frac{\partial \pi_m}{\partial u} = -c_m(u - \bar{u}) - h_m = 0 \quad \Rightarrow u = \bar{u} - \frac{h_m}{c_m}. \tag{13}$$

$$\frac{\partial \pi_r}{\partial v} = -w - c_r(v - \bar{v}) - h_r = 0 \quad \Rightarrow v = \bar{v} - \frac{w + h_r}{c_r}. \tag{14}$$

$$\frac{\partial \pi_r}{\partial p} = a\varphi - 2pb\varphi - h_r b = 0 \quad \Rightarrow p = \frac{1}{2}\left[\frac{a}{b} - \frac{h_r}{\varphi}\right]. \tag{15}$$

The parameter φ plays a key role in determining the price inside the two different schemes.

Lemma 1. *By adopting WPC in the WPC scheme, since $\varphi = 1$, the selling price results higher than that of RSC.*

$$p_{static}^{WPC} = \frac{1}{2}\left(\frac{a}{b} - h_r\right) > p_{static}^{RSC} = \frac{1}{2}\left(\frac{a}{b} - \frac{h_r}{\varphi}\right). \tag{16}$$

Since, in RSC $\varphi < 1$, this regime is characterized by a price lower than the wholesale price contract, [21,31] described Equation (16) by investigating the advantages and disadvantages of the two coordination contracts, whereby RSC mitigates the double marginalization effect.

Since the parameter φ does not exist in Equations (13) and (14), the quantities produced and sold in RSC and WPC are equal. Moreover, while Equation (13) remains always identical independent of the contract scheme, Equation (14) changes with respect to the transferring price w. In the WPC, in fact, the transferring price w_0 is higher than that of RSC. The manufacturer reduces the transferring price from w_0 to w whenever the revenue received from the retailer compensates for the decreasing wholesale price. Similarly, the retailer accepts the RSC if and only if the economic benefits due to a lower transferring price overcome the lower revenue. In sum, the players shift from WPC to RSC whenever

$$\pi^m_{RSC_{Static}} > \pi^m_{WPC_{Static}} \text{ and } \pi^r_{RSC_{Static}} > \pi^r_{WPC_{Static}}. \tag{17}$$

Upon introducing the sharing parameter and adjusting the transferring price, the players' strategies change. The manufacturer decides the producing quantity and reduces the transferring price, whereas the retailer purchases higher quantities at a lower price and shares revenues with the manufacturer. According to Equation (17), adopting RSC, both players result economically better off.

Lemma 2. *In WPC, the transferring price is higher than that of RSC, and this generates differences in the quantity purchased.*

$$q^{WPC}_{static} = \bar{v} - \frac{w_0 + h_r}{c_r} < q^{RSC}_{static} = \bar{v} - \frac{w + h_r}{c_r}. \tag{18}$$

The difference in the quantity purchased depends on the difference in the transferring prices.

$$q^{RSC}_{static} - q^{WPC}_{static} = \frac{w_0 - w}{c_r}. \tag{19}$$

Substituting Equations (13)–(15) into Equations (1) and (2), we obtain the firms' profit functions in static settings.

$$\pi^m_{RSC_{static}} = \frac{(1-\varphi)}{4}\left(\frac{a^2}{b} - \frac{h_r^2 b}{\varphi^2}\right) + (w + h_m)\left(\bar{v} - \frac{w + h_r}{c_r}\right) + \frac{h_m^2}{2c_m} - h_m(\hat{X} + \bar{u}). \tag{20}$$

$$\pi^r_{RSC_{static}} = \frac{\varphi}{4}\left(\frac{a^2}{b} - \frac{h_r^2 b}{\varphi^2}\right) - (w + h_r)\left(\bar{v} - \frac{w + h_r}{2c_r}\right) - h_r\left[\hat{Y} - \frac{1}{2}\left(a + \frac{h_r b}{\varphi}\right)\right]. \tag{21}$$

Proposition 3. *Under the static setting, RSC is more profitable than WPC for both M and R.*

Proof. Proposition 3 holds because of Lemmas 1 and 2. In RSC, the lower transferring price and the higher quantity purchased generate higher total profits than those under a WPC. According to [19], RSC mitigates the double marginalization effect and makes the players economically better off. □

3.3. Comparison between Centralized and Decentralized Solution of the Static SC Game

This section compares the qualitative solutions of centralized and decentralized configurations in terms of profit. Since the RSC generates higher profit than WPC, the relationships between the profits of centralized and decentralized SCs finally result as follows:

$$\pi^c_{Static} \geq \pi^m_{RSC_{Static}} + \pi^r_{RSC_{Static}} \geq \pi^m_{WPC_{Static}} + \pi^r_{WPC_{Static}}. \tag{22}$$

In the static formulation of the game, the centralized SC generates higher profit than the decentralized one in the RSC scheme, while the decentralized SC in WPC always generates lower profits compared to those of the previous scenarios. Among all possible sharing parameter values, artificial intelligence suggests adopting a sharing parameter that makes all firms economically better off. Interestingly, the artificial intelligence embedded in blockchains does not allow firms to reach the same outcomes of the centralized solution. In particular, the condition under which the centralized setting produces higher profits than RSC is $\frac{bh_r^2}{4} + (h_m - h_r)\left(\frac{h_m - h_r}{2c_r}\right) > \frac{bh_r^2}{4\varphi} + (h_m - h_r)\left(\frac{w - h_r}{2c_r}\right) - \frac{(w - h_r)^2}{2c_r}$. This condition highlights the importance of the parameter φ chosen by artificial intelligence for comparing the two settings. φ makes the RSC setting more convenient than the centralized one and, ceteris paribus, the previous condition does not hold. Considering $w = h_m$, the players do not take advantages from a centralized SC and, moreover, the decentralized SC results more profitable. Nevertheless, considering the parameter $\varphi = 1$, that is, hypothesizing a WPC scheme, the convenience in implementing a centralized SC depends on the condition $w > h_m$. Whenever this latter condition holds, in fact, the centralized setting always makes a higher profit than the decentralized one. If $w = h_m$, the profits under centralized SC and WPC are equal. Moreover, this is true when $h_m - h_r > 0$. Whenever this difference is small enough, the previous condition becomes $\frac{bh_r^2}{4} > \frac{bh_r^2}{4\varphi} - \frac{(w - h_r)^2}{2c_r}$; hence, the centralized setting is not suitable whenever $\varphi > \frac{bh_r^2}{bh_r^2 + 2\frac{(w - h_r)^2}{c_r}}$. Therefore, the artificial intelligence and the blockchain system should be created to use big data with the idea of achieving this shared target to make the supply chain better off from an economic and an operational point of view.

This paper extends the analysis of the game not only to the Nash solution but also to the Stackelberg one. This extension has the purpose of highlighting the convenience of playing either Nash or Stackelberg for improving the players' results as much as possible. Nevertheless, the structure of the game does not allow playing the Stackelberg game. When the manufacturer is the leader, the game does not change at all with respect to the Nash solutions. The manufacturer's decision variable, in fact, is u. As the Stackelberg game requires the integration of the leader's announced strategy inside the retailer's objective function, the game remains completely the same. The retailer's strategy is not influenced by the manufacturer's decisions. When the retailer is the leader of the game, again, the solution is not interesting at all. The retailer's decision variables are v and p, and both of them should be replaced inside the manufacturer's strategy. Nevertheless, neither v nor p influence the manufacturer's choices; thus, the Stackelberg game is not interesting as it coincides with the Nash solution. This result derives from the lack of interferences between the players' decision variables. The structure of the game matters when facing the comparison between Nash and Stackelberg games and should be adequately defined when the comparison is the aim of the study. In all these cases, the artificial intelligence system is not able to find an adequate contractual term to propose. The blockchain does not find a visible solution, and the miners are not able to solve the problem because the solution is equivalent to the previous one.

4. Characterization of the Dynamic Game

As for the static model proposed in Section 2, the purpose of the dynamic model is to maximize the firms' profit functions, $\pi_{Dynamic}^m$ and $\pi_{Dynamic}^r$. The differential game has the same structure as the static one; this similarity allows us to compare them appropriately. Also, in dynamic settings, in fact, the retailer decides the market price and purchased quantity given by p and v, respectively, while the manufacturer decides the production rate, u. The total revenue $[a - b(p)]\, p$ is shared with respect to the parameter $0 < \varphi < 1$ as the players adopt the RSC scheme. The artificial intelligence system is always in charge of deciding the optimal ranges for the sharing parameter. In fact, $\varphi = 1$ implies SC coordination through the WPC. The wholesale price and the sharing parameter are also determined by the blockchain, where the miners are asked to solve a more complicated algorithm, which is linked to the dynamics behind the differential games, whose outcomes would eventually lead to smart contracts.

In addition to the previous formulations, we introduce a digitalization option made through information sharing on inventory at all SC levels. This is realized through the adoption of Industry 4.0 technologies in the supply chain [24]. For describing the cost functions, the considerations made for the static game are still valid. The holding cost functions are assumed linear into the objective functions since their motions are controlled by [21] state equations.

$$\dot{X} = u - v \qquad X(0) = X_0 \geq 0. \qquad (23)$$

$$\dot{Y} = v - f(p) \qquad Y(0) = Y_0 \geq 0. \qquad (24)$$

By using these two state equations, this research investigates how players' stocks vary over time. In particular, Equation (23) represents the manufacturer's state equation, and shows how stocks are controlled over time by the manufacturer's production rate, u, and by the retailer's purchasing rate, v. Equation (24) is the retailer's state equation, which controls the retailer's stocks over time by the purchasing rate, v, and by the inverse demand function, $a - b(p)$.

The firm's profit functions are as follows:

$$\max_{u>0} \pi^m = \int_0^T e^{-rt}\{(1-\varphi)(a-bp)p + wv - c_m \frac{(u-\bar{u})^2}{2} - h_m X\}dt - \omega X e^{-rt}, \qquad (25)$$

$$\max_{v,p>0} \pi^r = \int_0^T e^{-rt}\{\varphi(a-bp)p - wv - c_r \frac{(v-\bar{v})^2}{2} - h_r Y\}dt - \sigma Y e^{-rt}. \qquad (26)$$

Since a short planning horizon is hypothesized, the discounting factor is not considered in our analysis.

4.1. Centralized Solution of the Dynamic SC Game

As for the static model, this paper presents the centralized dynamic model. This latter model produces a Pareto dynamic optimal solution, since both players make the highest possible profits. Whatever the coordination scheme adopted, the total profits generated are always lower or at least equal to the Pareto dynamic solution, which represents the benchmark of the game. Dynamic centralized settings were extensively proposed recently [27]. In this paper, the authors highlighted the benefits of collaboration in decentralized settings. In this paper, the artificial intelligence working in blockchains uses the centralized solution as a benchmark, searching for a sharing parameter that dynamically optimizes the firms' profits.

The centralized dynamic model is derived by summing up Equations (25) and (26). Since the players behave as a unique entity, the sharing parameter and the transferring price disappear.

$$\pi^c = \int_0^T \left[(a-bp)p - c_m \frac{(u-\bar{u})^2}{2} - c_r \frac{(v-\bar{v})^2}{2} - h_m X - h_r Y\right] dt - \omega X e^{-rT} - \sigma Y e^{-rT}. \qquad (27)$$

Also, in centralized settings, the discount factor is not considered since the time horizon is assumed to be short enough. By using Equations (23) and (24), the related Hamiltonian results as follows:

$$H^c = (a-bp)p - c_m \frac{(u-\bar{u})^2}{2} - c_r \frac{(v-\bar{v})^2}{2} - h_m X - h_r Y + \lambda_1^c (u-v) + \lambda_2^c [v - (a-bp)], \qquad (28)$$

where λ_1^d and λ_2^d are the co-state variables. Beyond including the state Equations (23) and (24), this problem considers the following constraints: At any instance of time, the feasible controls u and v

take non-negative values and the state constraints $X > 0$ and $Y > 0$ must be satisfied. Following this constraint, the necessary optimal conditions are derived for u, v, and p as follows:

$$H_u^c = -c_m(u - \bar{u}) + \lambda_1^c = 0 \quad \Rightarrow \quad u = \bar{u} + \frac{\lambda_1^c}{c_m}, \tag{29}$$

$$\dot{\lambda}_1^c = h_m \quad \lambda_1^c(T) = -\omega, \tag{30}$$

$$H_v^c = -c_r(v - \bar{v}) - \lambda_1^c + \lambda_2^c = 0 \quad \Rightarrow \quad v = \bar{v} + \frac{\lambda_2^c - \lambda_1^c}{c_r}, \tag{31}$$

$$\dot{\lambda}_2^c = h_r \quad \lambda_2^c(T) = -\sigma, \tag{32}$$

$$H_p^c = a - 2bp + \lambda_2^c b = 0 \quad \Rightarrow \quad p = \frac{1}{2}\left[\frac{a}{b} + \lambda_2^c\right]. \tag{33}$$

By using Equation (29) and the transversality condition in Equation (30), it is easy to show that $\lambda_1^c(t) < 0$ for $\forall t < T : t \in [0, T)$. $\lambda_1^c(t)$ reflects the implicit value of the inventory $X(t)$, and its negative value shows inefficiency in retaining or marginally increasing the stocks. This result is a consequence of the assumption $X > 0$. Centralized SC has no incentive to increase inventory. Equation (29) represents the optimal necessary condition for implementing an efficient producing policy in centralized SC. By differentiating $\dot{\lambda}_1^c$ and by using the relative transversality condition $\lambda_1^c(T) = -\omega$, it follows that $\lambda_1^c(t) = -\omega - h_m(T-t)$. Moreover, $u(t) < \bar{u} - \frac{\omega + h_m(T-t)}{c_m} \forall t < T : t \in [0, T)$ and $u(T) = \bar{u} - \frac{\omega}{c_m}$. This latter expression indicates that the centralized SC does not reach the optimal efficient level of production $\bar{u}$ at the end of the planning horizon. $-\frac{\omega}{c_m}$ represents the loss of efficiency as the salvage value is negative.

Similarly, by using Equation (31) and the transversality condition in Equation (32), as it is assumed that $Y > 0$, $\lambda_2^c(t) < 0$ for $\forall t < T : t \in [0, T)$. $\lambda_2^c(t)$ reflects the implicit value of the inventory $Y(t)$, and its negative value shows inefficiency in retaining or marginally increasing stocks. Also, the centralized SC has no incentive to increase inventory. Equation (31) represents the optimal necessary condition for implementing an efficient purchasing policy under centralized SC. By differentiating $\dot{\lambda}_2^c$ and by using the relative transversality condition $\lambda_2^c(T) = -\sigma$, it follows that $\lambda_2^c(t) = -\sigma - h_r(T-t)$. As $v(t) < \bar{v} - \frac{\sigma + h_r(T-t) - \omega - h_m(T-t)}{c_r} \forall t < T : t \in [0, T)$ and $v(T) = \bar{v} - \frac{\sigma - \omega}{c_r}$, the centralized SC misses the optimal efficient level of purchasing $\bar{v}$ at the end of the planning horizon. $-\frac{\sigma - \omega}{c_r}$ represents the loss of efficiency as the salvage value is negative.

From Equation (33), the price policy in centralized SC shows two critical elements values; $\frac{a}{b} > 0$ indicates the maximum market price, while λ_2^c is the marginal convenience in keeping stocks; hence, it has negative value.

In the centralized SC, the parameter φ disappears completely. The two players behave as a unique entity, taking advantage of centralization, while also eliminating the double marginalization effect. The transferring price is equal to the marginal producing cost. We use Equations (29), (31), and (33) to resolve the state equation in centralized SC.

$$\dot{X} = \bar{u} - \frac{\omega + h_m(T-t)}{c_m} - \bar{v} + \frac{\sigma + h_r(T-t) - \omega - h_m(T-t)}{c_r} \qquad X(0) = X_0 \geq 0. \tag{34}$$

$$\dot{Y} = \bar{v} - \frac{\sigma + h_r(T-t) - \omega - h_m(T-t)}{c_r} - \frac{a}{2} - \frac{b}{2}[\sigma + h_r(T-t)] \qquad Y(0) = Y_0 \geq 0. \tag{35}$$

By manipulating Equations (34) and (35) and differentiating the state equations with respect to time, we obtain the following system:

$$X = \left(\bar{u} - \bar{v} - \frac{\omega}{c_m} + \frac{\sigma - \omega}{c_r}\right)t + \left(\frac{h_r - h_m}{c_r} - \frac{h_m}{c_m}\right)\left(T - \frac{t}{2}\right)t + X_0^c, \tag{36}$$

$$Y = \bar{v} - \frac{a}{2} - \frac{\sigma - \omega}{c_r} - \frac{b\sigma}{2} - \frac{h_r(T-t) - h_m(T-t)}{c_r} + \frac{bh_r(T-t)}{2} + Y_o^c. \tag{37}$$

Substituting the values from Equations (29), (31), (33), (36), and (37) into Equation (27), the centralized SC profit function in dynamic settings is as follows:

$$\pi^c_{Dynamic} = \int_0^T e^{-rt}\left[\left(\frac{a^2}{b} - b(\lambda_2^r)^2\right) - \frac{(\lambda_1^c)^2}{2c_m} - \frac{(\lambda_2^c - \lambda_1^c)^2}{2c_r} - h_m X(t) - h_r Y(t)\right]dt - \omega X(t)e^{-rt} - \sigma Y(t)e^{-rt}. \tag{38}$$

4.2. Decentralized Solution of the Dynamic SC Game

In order to characterize the decentralized solution of the dynamic SC game, this session develops the Hamiltonian for both players showing necessary and sufficient conditions for optimality. The manufacturer's problem for determining the optimality conditions is as follows:

$$H^m = (1-\varphi)(a-bp)p + wv - c_m \frac{(u-\bar{u})^2}{2} - h_m X + \lambda_1^m(u-v) + \lambda_2^m[(v-f(p)], \tag{39}$$

where λ_1^m and λ_2^m are the co-state variables. The manufacturer's problem includes the two state Equations (23) and (24) and the following constraints: At any instant of time, the feasible control u takes non-negative values and the state constraints $X > 0, Y > 0$ must be satisfied.

The necessary optimality conditions are as follows:

$$H_u^m = -c_m(u - \bar{u}) + \lambda_1^m = 0 \quad \Rightarrow \quad u = \bar{u} + \frac{\lambda_1^m}{c_m}, \tag{40}$$

$$\dot{\lambda}_1^m = h_m \quad \lambda_1^m(T) = -\omega. \tag{41}$$

Because of the linearity of the game, the condition in Equation (40) is sufficient for optimality. Using Equation (41), we can show $\lambda_1^m(t) < 0$ for $\forall t < T : t \in [0, T)$. Therefore, from Equation (40) and the transversality condition in Equation (41), it follows that $u(t) < \bar{u}$ holds for $\forall t : t < T$, and $u(T) = \bar{u}$. $\lambda_1^m(t)$ reflects the manufacturer's marginal inventory cost and influences the value of X. Since it is negative, the manufacturer has no incentive to marginally increase stocks. $-c_m(u - \bar{u})$ reflects the economies of scale in production; as u approaches $\bar{u}$, the production cost decreases. $\lambda_2^m(t) = 0$ implies that the manufacturer does not influence the retailer's inventory policy.

The Hamiltonian maximization condition in Equation (41) shows that the marginal production cost is equal to the marginal inventory cost. Equation (40) represents the manufacturers optimal production policy. By differentiating Equation (41) and by using the transversality condition in Equation (21), it results that

$$\lambda_1^m(t) = -\omega - h_m(T-t). \tag{42}$$

Finally, the condition $u(t) < \bar{u}$ holds for $\forall t < T : t \in [0, T)$, and $u(T) = \bar{u} - \omega$; therefore, the optimal production rate is given by

$$u_* = \bar{u} - \frac{\omega + h_m(T-t)}{c_m}. \tag{43}$$

Similarly, the retailer's optimal control problem consists of determining the optimality conditions for v and p, which are the purchasing rate and market price, respectively.

$$H^r = \varphi(a-bp)p - wv - c_r \frac{(v-\bar{v})^2}{2} - h_r Y + \lambda_1^r(u-v) + \lambda_2^r[(v-f(p)], \tag{44}$$

where λ_1^r and λ_2^r are the retailer's co-state variables.

The retailer's problem includes the two state Equations (23) and (24) and the following constraints: At any instant of time, the feasible control v and p take non-negative values and the state constraints $X > 0$, $Y > 0$ must be satisfied.

Since we modeled a linear game, open- and closed-loop solutions coincide. The equilibrium degenerates at a closed-loop solution depending only on time; the final result is a sub-perfect Nash equilibrium, and the solution is Pareto optimal. Necessary optimality conditions for v and p are as follows:

$$H_v^r = -w - c_r(v - \bar{v}) - \lambda_1^r + \lambda_2^r = 0 \quad \Rightarrow \quad v = \bar{v} + \frac{\lambda_2^r - w}{c_r}, \tag{45}$$

$$\dot{\lambda}_2^r = h_r \qquad \lambda_2^r(T) = -\sigma, \tag{46}$$

$$\dot{\lambda}_1^r = 0 \qquad \lambda_1^r(T) = 0, \ \lambda_1^r(t) = 0 \qquad \forall t, \tag{47}$$

$$H_p^r = \varphi(a - bp) - \varphi bp + \lambda_2^r b = 0 \quad \Rightarrow \quad p = \frac{1}{2}\left[\frac{a}{b} + \frac{\lambda_2^r}{\varphi}\right]. \tag{48}$$

Equation (47) implies that the retailer does not influence the manufacturer's inventory policy. Equation (47) shows that $v(t) < 0$ for $\forall t < T : t \in [0, T)$, and, from Equation (21), it follows that $v(t) < \bar{v}$ for $\forall t < T : t \in [0, T)$ holds. $\lambda_2^r(t)$ reflects the retailer's marginal inventory cost and, implicitly, the value of Y. Negativity of this value implies that the retailer has no incentive to marginally increase stocks, according to the assumption $Y > 0$.

Equation (45) represents the retailer's optimal purchasing policy. By differentiating Equation (46), it results that

$$\lambda_2^r = -\sigma - h_r(T - t). \tag{49}$$

Therefore, $v(t) < \bar{v}$ for $\forall t < T : t \in [0, T)$ and $v* = \bar{v} - \frac{\sigma + h_r(T-t) + w}{c_r}$. At the end of the planning horizon, the optimal retailer's quantity is given by

$$v*(T) = \bar{v} - \frac{\sigma + w}{c_r}. \tag{50}$$

Equation (24) represents the necessary condition with respect to the price. In the retailer's price strategy, two elements appear; the first is $\frac{a}{b} > 0$, which indicates the maximum market price, and is strictly positive, while the second term $\frac{\lambda_2^r}{\varphi}$ is the retailer's marginal convenience in holding stocks, and its value is negative.

From this result, the main intuition concerns the double marginalization effect deriving from the transferring price and the marginal processing cost. According to [31], the double marginalization effect disappears as the transferring price is equal to the marginal production cost; all coordination contracts also contain, among others, this condition. Previous studies introduced several coordination schemes. Also, in the dynamic framework, the RSC mitigates the double marginalization effect. Since it contains theoretical and practical advantages and disadvantages, its results are compared with the WPC, essentially based on a fixed transferring price charged without considering any players' agreements, concerning neither revenue sharing nor price discounts. In particular, Equations (25) and (26) are structured in the RSC framework since the retailer shares some revenue defined by the parameter $\varphi : 0 < \varphi < 1$.

Lemma 3. *By adopting the WPC scheme in a dynamic framework, the selling price is higher than that of RSC.*

$$p_{dynamic}^{WPC} = \frac{1}{2}\left[\frac{a}{b} + \lambda_r^2\right] > p_{dynamic}^{RSC} = \frac{1}{2}\left[\frac{a}{b} + \frac{\lambda_r^2}{\varphi}\right]. \tag{51}$$

Inside the dynamic setting, the selling price increases over time, while both prices are equal at the end of the planning horizon.

Since $T > t$, $\lambda_2^r = -\sigma - h_r(T-t)$ is always decreasing and the price increases over time $\forall t < T : t \in [0,T)$. The lower φ is, the lower the price is. At the beginning of the planning horizon, $p^*(0) = \frac{1}{2}\left[\frac{a}{b} - \frac{\sigma + h_r(T)}{\varphi}\right]$; since $T > t$, the selling price increases over time for $\forall t : t \in [0,T)$ and converges to $p^*(T) = \frac{1}{2}\left[\frac{a}{b} - \frac{\sigma}{\varphi}\right]$ at the end of the planning horizon.

Nevertheless, the sharing parameter does not enter Equations (45) and (40); hence, the quantity produced and sold under RSC and WPC is identical. However, while Equation (40) is totally independent of the contract scheme since the transferring price does not appear, the difference in the transferring prices in the two contract schemes also generates a difference in the retailer's purchasing policy. In particular, the quantity purchased in WPC is lower than that of RSC, resulting in the following equation:

$$q_{dynamic}^{WPC} = \bar{v} - \frac{w_0 + h_r(T-t) + \sigma}{c_r} < q_{dynamic}^{RSC} = \bar{v} - \frac{w + h_r(T-t) + \sigma}{c_r}. \tag{52}$$

Since the transferring price does not enter Equation (40), similarly to the static game, the manufacturer does not modify its producing policy depending on the contract scheme. In contrast, the retailer changes his purchasing policy according to the difference between w and w_0. When coordinating the SC by RSC, the quantity purchased is higher. As the gap between transferring price increases, so does the distance between WPC and RSC quantities.

$$q_{dynamic}^{RSC} - q_{dynamic}^{WPC} = \frac{w_0 - w}{c_r}. \tag{53}$$

We can state the sufficient optimality conditions in terms of the Hamiltonian by using the Arrow approach. Consider the Hessian matrix for both players consisting of the four partial derivatives of the maximized Hamiltonian with respect to X and Y.

$$\mathbf{H}^m = \begin{bmatrix} H_{XX}^m & H_{XY}^m \\ H_{YX}^m & H_{YY}^m \end{bmatrix} \mathbf{H}^r = \begin{bmatrix} H_{XX}^r & H_{XY}^r \\ H_{YX}^r & H_{YY}^r \end{bmatrix}. \tag{54}$$

Proposition 4. *The necessary conditions for optimality are also sufficient.*

Proof. In the matrix in Equation (32), the off-diagonal elements are zero. Using the rule of principal minors, we conclude that, for both players, the maximized Hamiltonian is concave in (X, Y); hence, the necessary conditions for optimality are also sufficient. □

We use the results from Equations (40), (45), and (48) to resolve the state equations.

$$\dot{X} = \bar{u} - \frac{w + h_m(T-t)}{c_m} - \bar{v} + \frac{\sigma + h_r(T-t) + w}{c_r} \quad X(0) = X_0 \geq 0. \tag{55}$$

$$\dot{Y} = \bar{v} - \frac{\sigma + h_r^r(T-t) + w}{c_r} - \frac{a}{2} - \frac{b}{2}\left[\frac{\sigma + h_r(T-t)}{\varphi}\right] \quad Y(0) = Y_0 \geq 0. \tag{56}$$

By adequately manipulating Equations (55) and (56) and differentiating the state equations with respect to time, the solution of the previous system is as follows:

$$X(t) = \left(\bar{u} - \bar{v} - \frac{w}{c_m} + \frac{\sigma + w}{c_r}\right)t + \left(\frac{h_r}{c_r} - \frac{h_m}{c_m}\right)\left(T - \frac{t}{2}\right)t + X_0 = \alpha(t) + \beta\left(T - \frac{t}{2}\right)t + X_0, \tag{57}$$

$$Y(t) = \left(\bar{v} - \frac{\sigma + w}{c_r} - \frac{a}{2} - \frac{b\sigma}{2\varphi}\right)t - \left(\frac{h_r}{c_r} - \frac{bh_r}{2\varphi}\right)\left(T - \frac{t}{2}\right)t + Y_0 = \gamma(t) - \delta\left(T - \frac{t}{2}\right)t + Y. \tag{58}$$

Substituting the values from Equations (40), (45), (48), (57), and (58) into Equations (25) and (26), the firms' profit functions in dynamic settings are as follows:

$$\pi^m_{RSC_{Dynamic}} = \int_0^T e^{-rt}\left[\frac{(1-\varphi)}{4}\left(a^2 - \frac{b\lambda_2^{r\,2}}{\varphi^2}\right) + w\left(\bar{v} - \frac{w-\lambda_2^r}{c_r}\right) - \frac{(\lambda_1^m)^2}{2c_m} - h_m X(t)\right]dt - wX(t)e^{-rt}, \quad (59)$$

$$\pi^r_{RSC_{Dynamic}} = \int_0^T e^{-rt}\left[\frac{\varphi}{4}\left(a^2 - \frac{b(\lambda_2^r)^2}{\varphi^2}\right) - w\left(\bar{v} - \frac{w-\lambda_2^r}{c_r}\right) - \frac{(\lambda_2^r - w)^2}{2c_r} - h_r Y(t)\right]dt - \sigma Y(t)e^{-rt}. \quad (60)$$

Proposition 5. *Under the static setting, RSC is more profitable than WPC for both firms.*

Proof. Proposition 6 holds because of Lemma 3 and Lemma 4. In RSC, the lower transferring price and the higher quantity purchased generate higher total profits than those of WPC in a dynamic setting. RSC mitigates the double marginalization effect and makes the players economically better off. □

4.3. Comparison between Centralized and Decentralized Solutions of the Dynamic SC Game

The previous equation fixes an economic benchmark in the dynamic setting. Independent of the coordination scheme adopted, the total SC profits do not exceed that benchmark. According to Proposition 6, the RSC scheme generates higher profits than WPC, and knowing the Pareto solution produces generally higher results, the comparison between dynamic schemes results in the following equation:

$$\pi^c_{Dynamic} \geq \pi^m_{RSC_{Dynamic}} + \pi^r_{RSC_{Dynamic}} \geq \pi^m_{WPC_{Dynamic}} + \pi^r_{WPC_{Dynamic}}. \quad (61)$$

Centralized SC generates higher profits as compared to decentralized solutions coordinated by RSC or WPC. This is true when

$$\int_0^T -\frac{b\lambda_2^{c\,2}}{4} - \frac{(\lambda_2^c - \lambda_1^c)^2}{2c_r}dt > \int_0^T -\frac{b\lambda_2^{r\,2}}{4\varphi} - \frac{(\lambda_2^r - w)^2}{2c_r}dt > \int_0^T -\frac{b\lambda_2^{r\,2}}{4} - \frac{(\lambda_2^r - w)^2}{2c_r}dt. \quad (62)$$

These expressions represent centralized SC and decentralized SC under RSC and WPC in dynamic settings. The left-hand side of the expression shows the convenience of centralizing the SC expressed in terms of lower costs as compared with the RSC and WPC solutions reported in the middle and right-hand sides of the expression, respectively. The sharing parameter φ and the difference between λ_1^c and w make the various settings different. As $\varphi = 1$, RSC and WPC generate the same profits, and the players agree on centralizing the SC when $w > \lambda_1^c$. This is also true whenever $0 < \varphi < 1$. As $w = \lambda_1^c$, centralized and decentralized SCs under WPC generate the same results, and Equation (61) does not hold since the RSC produces lower profits. When $w = \lambda_1^c$ the manufacturer is indifferent with respect to producing and keeping stocks. He is not advantaged upon coordinating the SC; therefore, the simple WPC is preferable. When $w > \lambda_1^c$, centralization makes the players better off with respect to non-centralized solution; the manufacturer's inventory cost is in fact lower; hence, centralization produces efficiency in the higher part of the SC. By resolving the left-hand and middle parts of Equation (62), and by integrating and simply manipulating the equations, the threshold of the sharing parameter in dynamic setting results in the following equation:

$$\varphi > \frac{c_r b\left(bT + 2\sigma h_r \frac{T^2}{2} + h_r^2 \frac{T^3}{3}\right)}{\chi T + 2\varphi \frac{T^2}{2} + \eta \frac{T^3}{3}}, \quad (63)$$

with

$$\chi = c_r b^2 + 2(\omega - \sigma)^2 - 2(\sigma + w)^2$$
$$\varphi = c_r b \sigma h_r + 2(\sigma h_r - \sigma \omega - h_r \omega) - 2h_r(\sigma + w) \ . \quad (64)$$
$$\eta = c_r b h_r^2 + 2(h_m - h_r)^2 - 2h_r^2$$

As for the static setting, the dynamic development of the game does not consider the Stackelberg solution. This choice derives from the same motivations expressed previously. When assuming the manufacturer is the leader, upon announcing his decisions at first, the value of the decision variable u should be replaced inside the retailer's profit function in Equation (40), and then deriving the optimal conditions. As v and p are the retailer's decisions variables, the players' problems do not change. When assuming the retailer is the leader, upon announcing his strategy at first, the values of p and v should be replaced inside the manufacturer's profit function in Equation (25). Even in this case, the manufacturer's strategy does not change as p and v do not interfere with u. The decisions of each player are not influenced by the others' strategies; hence, the Stackelberg game does not provide more information than the Nash game. Once again, this result comes from the structure of the game. As our purpose is to compare the economical convenience in adopting alternative coordination schemes under static and dynamic settings, the next section develops the numerical analysis.

5. Numerical Analysis

Figure 2 summarizes all possible configurations derived previously considering two contract schemes (RSC and WPC), two settings (static and dynamic), and two configurations (centralized and decentralized SC).

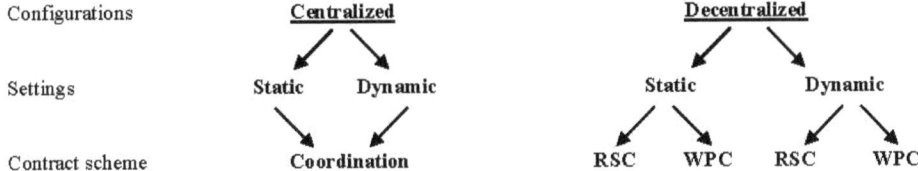

Figure 2. Possible combinations considering configurations, settings, and contract schemes.

In order to introduce a sound comparison between the previous models and then confirm and/or disconfirm existing results in the literature, we set up numerical analysis following the current trend of research in supply chain management using simulation models in this framework. Table 2 reports the base parameter values used.

Table 2. Base parameters.

a	b	φ	h_m	h_r	$\bar{u}$	$\bar{v}$	c_m	c_r	p_{WPC}	p_{RSC}	T
1000	20	0.6	5	5	100	100	5	5	70	5	100

With this setting, this study made the following assumptions:

1. The marginal production and purchasing costs, as well as the inventory costs, are the same; therefore, each player is indifferent with respect to producing or holding stocks.
2. Nevertheless, by producing and purchasing, each player can reach the optimal production and purchasing quantity. This represents the optimal amount of goods to attain in order to exploit the economies of scale and minimizing the total production and purchasing costs. Also, those quantities are assumed equal for both players.

3. The transferring price under RSC is equal to the production cost. The manufacturer does not increase his profit directly by selling but receives a compensating quota of the retailer's income. The parameter φ describes that proportion.

Figures 3–5 below were derived by solving the static and the dynamic games optimally and analyzing the cumulative profits. Solving the dynamic games allows one to sum up the cumulative profits over time of both firms by taking into consideration the constraints linked to the state variables along the co-state variables. The outcomes of the static games were computed by solving the game in one period and multiplying the profits for the same planning horizon considered in the dynamic settings. The findings of the research are aligned with the conclusions of [32], who found the application of a theoretical game very appealing and representative of reality. They invited researchers to use theoretical tools by calibrating the parameters according to the framework.

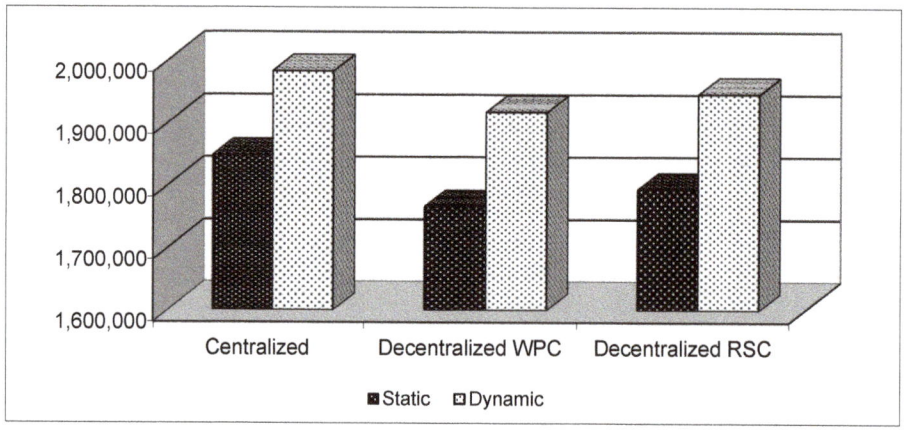

Figure 3. Comparison of cumulative profits between settings, contract schemes, and configurations.

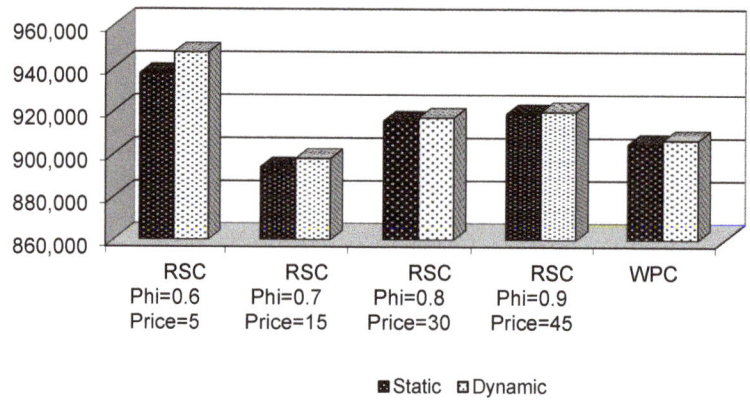

Figure 4. Sensitivity analysis of manufacturer's profits.

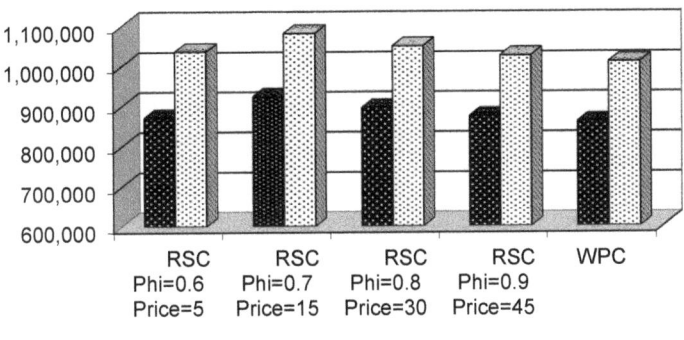

Figure 5. Sensitivity analysis of retailer's profits.

Figure 3 reports the comparison between the cumulative profits at the end of the planning horizon. Independent of the configuration and the contract scheme, the profits obtained under the dynamic setting were always higher than those resulting under the static one. This result introduces one important novelty in the literature showing which setting, static or dynamic, should be adopted when managing SCs. The dynamic setting always prevails in economic terms. It allows developing the economies of scale in production and purchasing and, therefore, achieving the optimal level of $\bar{u}$ and $\bar{v}$. Along the planning horizon, the production and purchasing rate become closer to those values, reducing the total production and purchasing costs at the end of the planning horizon. The two settings show equal gaps between actual and optimal production and purchasing level only in the beginning of the planning horizon. After that period, while the static setting still keeps the same gap until the end of the planning horizon, the dynamic one develops the economies of scale such that those costs decrease over time and confirm the statements introduced previously. Moreover, the dynamics related to the inventory policy optimize the cumulative profit over time. In the static setting, the inventories remain equal over the entire planning horizon resulting much higher that the dynamic one, i.e., not optimal at all. Finally, the dynamic setting shows an increasing price over time according to its dynamic motion. Under static settings, the price is always constant and, above all, is higher than the dynamic one. This result is always true independent of the coordination scheme and the configuration. Consequently, the demand related to the static settings is always lower than that obtainable under dynamic settings; therefore, it results in lower revenues and cumulative profits.

As confirmed by the flow of contributions in the literature, the centralized SC always results in higher cumulative profits than the decentralized one. In this sense, the centralized SC is a benchmark; whatever the coordination scheme adopted, its results can be at most attained according to the smart contract terms. The simultaneous comparison of contract schemes and configurations reveals that the RSC is highly economically attractive for coordinating decentralized SCs. Compared to the WPC, it reaches a similar level of profits to the centralized SC. This is true especially under dynamic settings. According to the literature, the RSC contract generates higher profits than the WPS. The results of this study partially confirm this previous statement. In particular, it remains valid as the comparison involves the same setting. The RSC under dynamic (static) settings makes higher cumulative profits than the WPC under dynamic (static) settings. This result is well known in the literature. One of the main novelties proposed by this research concerns the comparison of the two coordination schemes considering the setting as well. In particular, the WPC under dynamic settings generates higher cumulative profits than the WPC under a static framework. This result is quite new in the literature. Previous contributions compared the two coordination schemes by always evaluating the same setting. The choice of the setting matters consistently. In the case of a decentralized SC, the adoption of the most appropriate coordination scheme should include the selection of the right setting as well. By coordinating the SC through a WPC and working under a dynamic setting, the economic results improve significantly. The RSC fails in coordinating the SC under static settings when compared with the WPC under dynamic settings. This result finds confirmation from some theoretical results introduced by [19]. The RSC, in fact, is an effective coordination contract scheme; however, it is very expensive and difficult to administer. From this point of view, the WPC is always preferred because it is less problematic. Although RSC could appear extremely attractive, the higher part of the SC adopts WPC, losing some economic benefits expressed in terms of profit but saving time and cost for implementation and control. The results of this research show that, beyond losses and difficulties, the right choice of the coordination scheme also involves the choice of the setting; the RSC is more convenient than the WPC, while that statement contrasts with the results of this research when considering the setting.

Moreover, another important novelty concerns the comparison between decentralized SC under dynamic settings and centralized SC under a static framework. While the flow of contributions highlighted the supremacy of centralized SCs with respect to decentralized SCs, different results emerged when comparing the two settings. In particular, when coordinating SC by means of RSC under dynamic settings, the economic outcomes generated are higher than that of the centralized SC under a static framework. Additionally, the WPC under dynamic settings makes higher cumulative profits than the centralized SC under a static environment. Independent of the coordination scheme adopted, the decentralized SC under dynamic settings produces higher economic results compared to the centralized one under static settings. This result suggests a significant novelty inside the SC literature and introduces a controversial result with respect to the existing contributions. While, up to now, the literature proposed the centralized SC as a primary benchmark for comparing any other scenario, when evaluating configurations belonging to different settings, the results disconfirm part of the existing literature. In particular, this research addresses the appropriateness of investigating the SC management as a dynamic rather than a static phenomenon. In this sense, great merit goes to the use of artificial intelligence and blockchain systems. In fact, the artificial intelligence searches for optimal values of the sharing parameter that allow the firms' decentralized strategies to mimic the outcomes of a centralized solution. Surprisingly, this job allows firms to better understand the power of a dynamic supply chain in a digital economy. The economic performance of a decentralized supply chain is considerably higher than a static centralized approach, which disregards the use of an artificial intelligence system or blockchain, as well as any form of Industry 4.0 tool. Countless variables involved in SC coordination behave differently at different instants of time; therefore, the application of the static approach generates myopic results. The dynamic investigation allows appreciating the movements and the trajectories of some variables over time such that the players act by evaluating the performance and consequently managing SCs following those changes. The particular case proposed by this research shows that the benefits generated by dynamically evaluating production and purchasing costs, inventories, and demand create controversial results with respect to 20 years of study of SC management. The decentralized SC under dynamic settings is always preferred to the centralized SC under a static environment.

Beyond comparing the economical results from centralized and decentralized settings, this research compares the firm's cumulative profits under decentralized settings for several values of φ and w. While the base case reports precise values for both parameters, here, sensitivity analysis was applied in order to investigate whether the previous statements find confirmation in the relationships between firms under decentralized settings and to identify who benefits more when implementing a particular configuration. In particular, φ was assumed to take values comprised in the interval (0.6, 1], while w took values comprised in the range (5, 60). Interestingly, the artificial intelligence system proposes the use of a sharing parameter equal to 0.6, as well the minimum wholesale price. This combination can result very challenging for some companies, especially when working in a global supply chain. In fact, firms cannot control the amount of revenue that firms transfer in their transactions. Therefore, the use of blockchains helps substantially in better controlling the transaction, trusting the relationships, and making it visible to all suppliers. Finally, coordination is possible when artificial intelligence is used, as it uses the coordinated solution as a benchmark to derive the optimal contractual parameters. In this sense, we highlight that the use of blockchains and artificial intelligence for accounting duties only is a useless investment. These digital technologies should instead be used for strategic parameters like the sharing parameter and the wholesale price, requiring a high level of negotiation among suppliers.

When evaluating the effect of different combinations of φ and w on the manufacturer's cumulative profits, the finding reveals particular appreciation for the RSC in eliminating the double marginalization effect. As the manufacturer sells the products at the minimum selling price that is equal to the marginal cost ($w = c_m = 5$) and obtains a high portion of the retailer's revenue, the supply chain takes a configuration of RSC able to mitigate the double marginalization effect and substantially increase the manufacturer's profits. Figure 4 reports the sensitivity analysis of the two parameters for appreciating how the cumulative profit varies and identifying the most convenient configuration. Comparing these parameters from the extreme RSC with $w = 5$ and $\varphi = 0.6$ to the WPC with $w = 60$ and $\varphi = 1$, this research shows that the RSC is preferred to the WPC in most cases. The elimination of the double marginalization effect generates better results for the manufacturer with respect to the WPC that gives lower cumulative profits. This statement is highly influenced by the couple of values (w, φ). For some combinations of values, the RSC always generates higher results than the WPC independent of static or dynamic settings. However, some other combinations of values disconfirm this statement. They reveal in fact that the WPC is preferred to the RSC and that the setting matters. Moreover, some combinations of parameters imply always preferring the WPC. The manufacturer should appropriately evaluate the two parameters as the economic effects change considerably.

Independent of the combination of (w, φ), the retailer's results appear quite stable. Although the RSC mitigates the double marginalization effect, that contract generates a lower benefit for the retailer. In particular, starting from the extreme configuration $w = 5$ and $\varphi = 0.6$ and considering several possible combinations, the cumulative profit does not increase as significantly as the manufacturer's profit. When evaluating the profits within the same setting, the results of this research confirm the existing theory in the theme of SC coordination and contract; the RSC generates higher economical results that the WPC as it is able to mitigate the double marginalization effect. Nevertheless, this well-assessed statement is valid whenever the evaluation of two contract configurations involves the same setting [33]. From Figure 5, it appears quite clear that all configurations of RSC under dynamic (static) settings generate higher results than the WPC under dynamic (static) settings. Notwithstanding, the results vary consistently when comparing heterogeneous settings. In particular, according to the results obtained previously when comparing centralized and decentralized solutions, the WPC under dynamic settings generates better results than the WPC under a static environment. Consequently, the results of this research confirm the previous findings in the literature showing the supremacy of the RSC with respect to the WPC. Since the main purpose of this research concerns the comparison of static and dynamic approaches for coordinating the SC, the results disconfirm part of the existing literature. In particular, it addresses the importance of the setting when choosing the more appropriate contract scheme. In particular, for the retailer, the coordination through the RSC under static settings appears less convenient than the WPC under a dynamic environment. The RSC is difficult to administer and implement, and it does not bring more advantages than the WPC when evaluating them by considering the settings as well. The use of blockchains and artificial intelligence will surely make firms more comfortable in using complex contracts like RSC in the future.

6. Conclusions

Inspired by the contributions by the recent digital transformation, this research developed a SC game including marketing and dynamic inventory decisions. The main novelties of the paper concern the implementation of the SC game under dynamic and static settings, as well as the comparison of two contract schemes, wholesale price contract and revenue sharing contract, used for SC coordination. An artificial intelligence system supports the decision-making process and determines the optimal contract terms. In particular, the artificial intelligence uses the centralized solutions to identify the best contract clauses to be used by the firms. The blockchain allows the complexity of smart contracts to be managed such that firms can easily use digitalization in their transactions.

Our results are aligned with the literature comparing centralized and decentralized SCs within the same setting; the centralized SC under dynamic (static) settings generates higher profits than the decentralized SC under dynamic (static) settings. This result was obtained from the qualitative analysis and within the numerical resolution, and it was valid independent of the coordination scheme adopted (RSC or WPC). Nevertheless, the evaluation of the setting matters substantially. We are able to provide the following findings:

1. While the existing contributions successfully assessed that the RSC is preferred to the WPC for coordinating SCs, this statement fails when comparing the WPC under dynamic settings with the RSC under static settings. In particular, the cumulative profits obtained by using WPC under dynamic settings result higher than those generated by implementing RSC under a static environment. Accordingly, SCs should be coordinated by simultaneously evaluating the contract schemes and the setting and converging toward an optimal decision. An artificial intelligence system, along with blockchain and big data, should be implemented according to these targets rather than as mere smart tools to write lines of orders.

2. Existing contributions clearly highlight the preference for the centralized SC to the decentralized one. We compare the cumulative profits finding that the decentralized SC under dynamic settings obtains higher profits than those obtained by the centralized SC under a static framework. This statement is true independent of the contract scheme adopted for SC coordination. This result suggests that the decision-maker cannot disregard the setting when choosing the SC configuration. Static and dynamic settings suggest an important innovation in the literature when evaluating centralized and decentralized SC compositions. Figure 6 reports the summary of our findings showing the convenience when going from one configuration to another. The bold arrows reflect the innovation due to this research, while the others concern the results already known and well established in the literature and confirmed here.

3. The choice of the sharing parameter φ and the transferring price w determines the convenience for each player in adopting one configuration rather than another. This is particularly true for the manufacturer. When φ is low enough and the transferring price is equal to the marginal production cost, the RSC totally mitigates the double marginalization effect, and it is found to always be highly preferred with respect to the WPC. When the values (φ, w) change, the results are not obvious. The smart contracts that firms use should aim at searching for the optimal combinations of these two parameters to make SCs better off with digitalization. Our findings suggest that, beyond the choice of setting, the adequate combination of the parameters (φ, w) plays an important role in choosing the optimal configuration for maximizing the manufacturer's cumulative profit.

4. Finally, the choice of the parameters (φ, w) substantially influences the retailer's cumulative profits. Nevertheless, the retailer shows more stable and interesting results with respect to the comparison between static and dynamic settings. The implementation of the RSC is always preferred to the WPC for the higher cumulative profits generated by each combination of (φ, w). Notwithstanding, this result is true as long as the comparison between the two contract schemes uses the same settings. When evaluating the results of the WPC under dynamic settings with the RSC under static settings, the previous statement fails for each combination of (φ, w) used in the sensitivity analysis. The retailer incurs higher economic benefits by adopting WPC in dynamic settings than RSC in static settings. This result introduces a novelty in the literature. When coordinating the SC, the choice of the setting matters considerably, and the adoption of the contract scheme depends on the selection of the setting. Figure 7 reports the summary of the firms' convenience in shifting from one configuration to another. While the results concerning the retailer are quite stable and clear, the choice of the parameters (φ, w) impacts the manufacturer's

decisions, which is the leading firm in terms of implementing an artificial intelligence system. Non-bold arrows illustrate the well-assessed findings in the literature, while the findings of this paper are shown in bold. Bold and double arrows represent the relationships that need further future investigations, as the results obtained are not at all definitive.

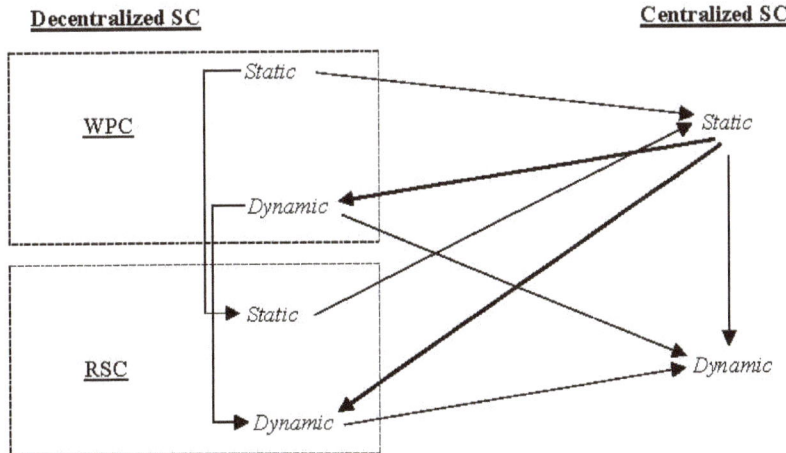

Figure 6. Summary of the comparison of centralized and decentralized supply chains (SCs).

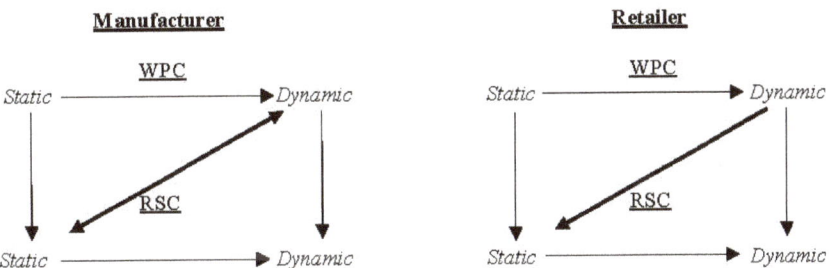

Figure 7. Summary of the comparison of manufacturers and retailers.

This paper is not free of limitations, which are listed to inspire future research in this area. This research most likely focuses on operational conditions. Future work can analyze the benefits of artificial intelligence and blockchains, along with smart contracts, in other contexts, such as multi-channel, omni-channel, and closed-loop SCs [34]. Future research can look to obtain real data to validate the model rather than using simulated data. Other technologies can also be evaluated in SC frameworks, such as big data, three-dimensional (3D) printing, and cloud computing.

Funding: This research received no external funding.

Conflicts of Interest: The authors declare no conflict of interest.

References

1. Mentzer, J.T.; DeWitt, W.; Keebler, J.S.; Min, S.; Nix, N.W.; Smith, C.D.; Zacharia, Z.G. Defining supply chain management. *J. Bus. Logist.* **2001**, *1*, 1–25. [CrossRef]
2. Kogan, K.; Tapiero, C. *Supply Chain Games: Operations Management and Risk Evaluation*; Springer Series in Operation Research; Springer: Berlin/Heidelberg, Germany, 2007.

3. De Giovanni, P.; Ramani, V. Product cannibalization and the effect of a service strategy. *J. Oper. Res. Soc.* **2018**, *69*, 340–357. [CrossRef]
4. Ramani, V.; De Giovanni, P. A two-period model of product cannibalization in an atypical Closed-loop Supply Chain with endogenous returns: The case of DellReconnect. *Eur. J. Oper. Res.* **2017**, *262*, 1009–1027. [CrossRef]
5. Jeuland, A.P.; Shugan, S.M. Managing Channel Pro.ts. *Mark. Sci.* **1983**, *2*, 239–272. [CrossRef]
6. Jørgensen, S. Optimal production, purchasing and pricing: A differential game approach. *Eur. J. Oper. Res.* **1986**, *24*, 64–76. [CrossRef]
7. Sethi, S.P. Dynamic Optimal Control models in advertising: A survey. *SIAM Rev.* **1977**, *19*, 685–725. [CrossRef]
8. Feichtinger, G.; Hartel, R.F.; Sethi, S.P. Dynamic optimal control models in advertising: Recent developments. *Manag. Sci.* **1994**, *40*, 29–31. [CrossRef]
9. Dockner, E.; Jørgensen, S.; Long, N.V.; Sorger, G. *Differential Games in Economics and Management Science*; Cambridge University Press: Cambridge, UK, 2000.
10. Erickson, G.M. Differential game models of advertising competition. *Eur. J. Oper. Res.* **1995**, *83*, 431–438. [CrossRef]
11. Erickson, G.M. Empirical analysis of closed-loop duopoly advertising strategies. *Manag. Sci.* **1992**, *38*, 1732–1749. [CrossRef]
12. He, X.; Prasad, A.; Sethi, S.; Gutierrez, G. A Survey of Stackelberg Differential Game Models in Supply and Marketing Channels. *J. Syst. Sci. Syst. Eng.* **2007**, *16*, 385–413. [CrossRef]
13. Jørgensen, S.; Zaccour, G. *Differential Games in Marketing, International Series in Quantitative Marketing*; Kluwer Academic Publishers: Boston, UK, 2004.
14. Taboubi, S.; Zaccour, G. *Coordination Mechanisms in Marketing Channels: A Survey of Game Theory Models*; Cahier du GERAD, G-2005-36; GERAD: Montreal, QC, Canada, 2005.
15. De Giovanni, P. *Digital Supply Chain, Design Quality and Circular Economy, in Dynamic Supply Chain Quality Models in Digital Transformation*; Springer: Berlin/Heidelberg, Germany, 2020.
16. De Giovanni, P. A joint maximization incentive in closed-loop supply chains with competing retailers: The case of spent-battery recycling. *Eur. J. Oper. Res.* **2018**, *268*, 128–147. [CrossRef]
17. De Giovanni, P. Coordination in a distribution channel with decisions on the nature of incentives and share-dependency on pricing. *J. Oper. Res. Soc.* **2016**, *67*, 1034–1049. [CrossRef]
18. De Giovanni, P.; Zaccour, G. Optimal quality improvements and pricing strategies with active and passive product returns. *Omega* **2019**, *88*, 248–262. [CrossRef]
19. Cachon, G.; Netessine, S. Game theory in Supply Chain Analysis. In *Handbook of Quantitative Supply Chain Analysis: Modeling in the eBusiness Era*; David Simchi-Levi, S., Wu, D., Shen, Z., Eds.; Kluwer: Alphen aan den Rijn, The Netherlands, 2004.
20. Taylor, T.A. Supply chain coordination under channel rebates with sales effort effects. *Manag. Sci.* **2002**, *48*, 992–1007. [CrossRef]
21. Cachon, G.P.; Lariviere, M.A. Supply Chain coordination with revenues sharing contracts: Strength and limitations. *Manag. Sci.* **2005**, *51*, 30–44. [CrossRef]
22. Tsay, A.A. The quantity flexibility contract and supplier-customer incentives. *Manag. Sci.* **1999**, *45*, 1339–1358. [CrossRef]
23. De Giovanni, P. An optimal control model with defective products and goodwill damages. *Ann. Oper. Res.* **2019**, 1–12. Available online: https://link.springer.com/article/10.1007/s10479-019-03176-4 (accessed on 1 October 2019).
24. De Giovanni, P. Eco-Digital Supply Chain through Blockchains. In Proceedings of the 9th International Conference on Advanced in Social Science, Economics and Management Study, Rome, Italy, 7–8 December 2019.
25. Zaccour, G. On the Coordination of Dynamic Marketing Channels and Two-Part Tariffs. *Automatica* **2008**, *44*, 1233–1239. [CrossRef]
26. De Giovanni, P.; Karray, S.; Martín-Herrán, G. Vendor Management Inventory with consignment contracts and the benefits of cooperative advertising. *Eur. J. Oper. Res.* **2019**, *272*, 465–480. [CrossRef]
27. Liu, B.; De Giovanni, P. Green Process Innovation Through Industry 4.0 Technologies and Supply Chain Coordination. *Ann. Oper. Res.* **2020**, *1*, 1–32.

28. Wang, G.; Gunasekaran, A.; Ngai, E.W.; Papadopoulos, T. Big data analytics in logistics and supply chain management: Certain investigations for research and applications. *Int. J. Prod. Econ.* **2016**, *176*, 98–110. [CrossRef]
29. Jørgensen, S. A survey of some differential games in advertising. *J. Econ. Dyn. Control* **1982**, *4*, 341–369. [CrossRef]
30. Genc, T.S.; De Giovanni, P. Optimal return and rebate mechanism in a closed-loop supply chain game. *Eur. J. Oper. Res.* **2018**, *269*, 661–681. [CrossRef]
31. Spengler, J. Vertical integration and antitrust policy. *J. Political Econ.* **1950**, *58*, 347–352. [CrossRef]
32. Chalikias, M.; Skordoulis, M. Implementation of FW Lanchester's combat model in a supply chain in duopoly: The case of Coca-Cola and Pepsi in Greece. *Oper. Res.* **2017**, *17*, 737–745.
33. Preeker, T.; De Giovanni, P. Coordinating innovation projects with high tech suppliers through contracts. *Res. Policy* **2018**, *47*, 1161–1172. [CrossRef]
34. De Giovanni, P.; Zaccour, G. A selective survey of game-theoretic models of closed-loop supply chains. *4OR* **2019**, *17*, 1–44. [CrossRef]

© 2019 by the author. Licensee MDPI, Basel, Switzerland. This article is an open access article distributed under the terms and conditions of the Creative Commons Attribution (CC BY) license (http://creativecommons.org/licenses/by/4.0/).

Article

Impacts of Online and Offline Channel Structures on Two-Period Supply Chains with Strategic Consumers

Qian Lei, Juan He * and Fuling Huang

School of Transportation and Logistics, Southwest Jiaotong University, Chengdu 611756, China; leiqian900219@163.com (Q.L.); fumao723519@163.com (F.H.)
* Correspondence: hejuan@swjtu.edu.cn

Received: 12 September 2019; Accepted: 22 December 2019; Published: 27 December 2019

Abstract: In this paper, the effects of strategic consumer behaviors have been investigated and analyzed with regard to online retailers and offline retailers in a dual-channel supply chain. Four channel structures (i.e., no-promotion, a direct online channel, a retail offline channel, and dual channels introduced in the promotion sales period) are considered. At the beginning of the paper, the original demand functions of a dual-channel supply chain incorporating the consumers' utility has been introduced. The results indicate that despite improved consumer patience, all promotional prices do not fall as expected. When sales channels are provided by online retailers rather than offline retailers during the promotion period, offline retailers can achieve higher profits. We also find that in most cases, a dual-channel model in a single-period is more beneficial to both online and offline retailers than a dual-channel model in two periods, which is, to a certain extent, contrary to the existing literature of single sales channel.

Keywords: dual-channel; two sales periods; channel structure strategy; strategic consumers; Nash game

1. Introduction

The Internet has significantly influenced consumers' purchase patterns. Some consumers may prefer to purchase online, while others may prefer to shop in stores (i.e., offline). Consequently, different consumer purchase patterns have inspired online and offline retail channels (i.e., dual channels). According to eMarketer (2018), e-retail sales accounted for 10.2 percent of all retail sales worldwide in 2017 and expected to reach 17.5 percent in 2021. However, prices for products tend to be marked down after new versions of products are introduced into the market. Notably, about 50% of inventory is sold at discount prices in the clothing industry [1]. Purchasers of automobiles, home appliances, and other durable goods also routinely wait for prices to fall. There are several different kinds of channel structures in current marketing systems to sell overstocked products, such as the traditional offline retail only, the online only, and dual-channel promotion, which is a combination of the first two channels. According to Adobe Analytics Data, Cyber Monday, acting as the most classic online promotional activities in the second sales period, sales topped $7.9 billion in 2018. Black Friday is one of the representative offline promotional activities, and shoppers spent nearly $6.22 billion on the day, 23.6% up from last year. Another example is the Christmas or Chinese New Year, during which both the offline and online retailers intend to offer big discounts. These examples demonstrate the effectiveness of online and offline promotional strategies.

These observations motivated us to explore the impact of channel market structure on how an offline/online retailer makes strategic decisions to clear overstocked products in the second period. In this paper, a two-period model for the online and offline channels that sell the same products has been developed. In the first sales period, regular-priced products are sold in both the offline channel and the online channel at the same time, while in the second sales period, overstocked products are sold at discounted prices through three different channel structures: the direct online channel only (defined as "Model D"), or the retail offline channel only (defined as "Model R"), or both channels (defined as "Model B"). In addition, we also wondered whether the extended sales period is beneficial to the retailers by considering no promotion situation (defined as "Model N"). In each model, we are interested in investigating the effects of strategic consumers' behaviors on the whole system equilibrium.

We characterize the strategic consumers' behaviors, including the degree of consumer patience and the acceptance of the direct channel into two-period offline/online channel sales models. Whether consumers purchase during the first or second sales period mainly depends on the degree of consumer patience. Consumers may wait for the lowest possible discounted price before making a purchase [2]. According to the Market Research Society's survey, more than 50% of consumers tend to wait for the low-price period to buy products. In addition, the acceptance of the direct channel reflecting consumers' willingness to purchase online is a key factor. When strategic consumers confront dual channels, consumer acceptance of the direct channel may be less than the acceptance of conventional retail stores because delivery time exists in a direct channel.

We construct a Nash game-theoretic model to represent the interaction among the online retailer, the offline retailer, and a population of strategic consumers in the four above-mentioned channel market structures. The contribution of this study is threefold. First, demand functions at each sales period through offline or online channel are firstly introduced by incorporating the consumer's utility in our paper. Second, optimal quantity strategies for the online and offline retailers in four dual-channel supply chains are established, respectively. Third, the pricing decisions and profits of online and offline retailers among different sales channel structures are compared to judge whether offline or online channel in the second sales period should be offered or which model should be adopted.

We make some interesting observations. First, even though the degree of consumer patience increases, all selling prices in the promotion period set in Model D, Model R, and Model B do not decrease as expected. Meanwhile, the offline retailer's optimal quantities and prices do not always decrease with the increase in the acceptance of the online channel. In particular, the offline retailer's results in the promotion period increase with increasing acceptance in Model R. Second, the online retailer's profit and offline retailer's profit in Model B are lower than those in Model N, respectively, which is contrary to the results reported in the existing literature. Third, compared with introducing an offline channel in the promotion period by the offline retailer (i.e., Model R), the offline retailer could obtain higher profit when the online retailer sets the online channel in the promotion period (i.e., Model D).

The remainder of this paper is organized as follows. Section 2 reviews the relevant literature. Section 3 introduces the innovative demand functions of the four models described above. Section 4 derives the optimal decisions of online and offline retailers for the different retail structure strategies. Section 5 compares the equilibrium results of the different retail structure strategies by numerical examples. Section 6 concludes the results and outlines limitations. Finally, Appendix A presents the proofs for the Propositions.

2. Literature Review

Our paper focuses on the research of an offline/online dual channels supply chain. The dual-channel supply chain with a single-period model has been given much attention, mainly on channel choice. Arya et al. (2007) showed that the online channel plays an important role in exerting potential competition pressure on the existing retailer by increasing the manufacturer's negotiation power [3]. Moreover, Chiang et al. (2013) found that the introduction of an online channel always results in a wholesale price reduction, which might benefit both the retailer and the manufacturer [4]. Li et al. (2019) investigated the strategic effect of return policies in a dual-channel supply chain where the manufacturer decides whether to implement a return policy in the online channel, the offline channel, and dual channels [5]. However, only one single sales period is considered in these essays; we concentrate on the channel choice in the promotion sales period, which has seldom been touched. A considerable body of research also concentrates on pricing strategies in the dual-channel supply chain. Hua et al. (2010) examined the optimal decisions of a dual-channel model under the condition of the linear demand function, but they did not take strategic consumers into account [6]. Many other factors, such as delivery lead time [7], product availability for offline channel's service [8], return policy adoption [5,9], are considered to examine how these factors affect the whole pricing strategy of dual channels.

A significant amount of work has been carried out on the two-period model. De Giovanni and Zaccour (2014) investigated the pricing, collection effort decisions, and members' profitability to compare several two-period closed-loop supply chain configurations of the collection process [10]. Lin (2016) assumed that the demand function is linear and includes reference price effects, they found the reference price effect could alleviate the double marginalization effect and improve the channel efficiency [11]. In the same demand setting, Maiti and Giri (2017) developed four decision strategies, including the same or different wholesale prices to two selling periods in preannounced or delayed pricing strategies [12]. When the demand at each period is stochastic, Chen and Xiao (2016) investigated the optimal decisions of the players, where stock-out and holding costs are incorporated into the two-period model [13]. However, neither of these studies addressed price-dependent demands by introducing strategic consumers' behavior, such as consumers' patience. Papanastasiou and Savva (2017) touched upon the consumer degree of patience in the two-period supply chain [14]. But they ignored the strategic behaviors in the dual channels model.

By combining the dual-channel with a two-period supply chain, we observed studies on the two-period dual-channel supply chain, which are also closely related to this paper. For instance, Lai et al. (2010) investigated the price matching strategy, which eliminates the enthusiasm of consumers to delay buying, thus allowing retailers to increase prices during normal sale periods [15]. Huang et al. (2012) developed a two-period pricing and production decision model in a dual-channel supply chain that experiences demand disruption during the planning horizon. They showed that optimal prices are affected by consumers' channel preferences and the market scale [16]. Nevertheless, what we obtained reveals that the optimal pricing structure also depends on the degree of consumer patience. Xiong et al. (2012) considered a durable goods market consisting of direct sales by manufacturers through online and offline channels and a mix of selling and leasing by dealers through the offline channel to consumers [17]. In the same setting with [17], Yan et al. (2016) investigated how the addition of the online channel affects the traditional marketing strategies of leasing and selling [18]. The previous literature do not consider strategic consumer behaviors.

There has been a growing interest in studying the impact of strategic consumer behavior on a seller's pricing decision. Hübner et al. (2013) structured retail demand and supply chain planning questions coherently from the perspective of suppliers and consumers [19]. In Cachon and Swinney (2009), consumers may wait for a clearance sale, the probability of which is low if the seller can better match supply with demand using advance demand information [20]. In our paper, both the acceptance of the online channel and the degree of consumer patience for waiting for the second sales period are discussed in a dual-channel supply chain. Furthermore, strategic consumer behaviors are also discussed in other marketing settings. For example, Giampietri et al. (2018) investigated consumer's motivations and behaviors in a short food supply chain [21]. Zimon and Domingues (2018) developed guidelines for the concept of sustainable supply chain management in the textile industry [22]. Zhou (2016) focused on the pricing strategy of a dual sales channel where the price in the second period has no effect on the market demand in the first period [23]. By contrast, this effect is elaborated in our model. We also incorporate the effect of several channel structures on profits to provide insights for the choice of sales structure.

To the best of our knowledge, the supply chain with dual-channel, two sales periods, and strategic consumers described in this paper has seldom been studied. For the problem described above, we characterize the degree of consumer patience and the acceptance of the direct channel into a two-period dual-channel sales model. We also present a numerical analysis for choosing the optimal channel structural strategy.

3. Models

In this section, we describe the basic model of consumer choice and the demand functions within different channel structures over two periods in detail.

3.1. Online and Offline Retailers

This paper focuses on four common models, online and offline retailers would have four different quantitative behaviors. The first model is a benchmark model without considering the second promotion period (denoted "Model N"). In the second model (denoted "Model D"), only the online seller provides a price discount in the second period. This setting is motivated by the online promotion on "Cyber Monday" in the United States or "Double Eleven" in China. Inspired by store promotion on "Black Friday", some offline retailers promote products in the second period (denoted "Model R"). At last, both the retailers adopt promotion strategies in dual channels (denoted "Model B"). Figure 1 presents the sequence of events under the four models.

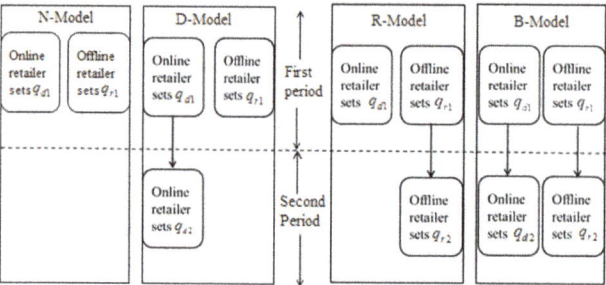

Figure 1. Graphical representation of decision sequences for four models.

The interactions between the online retailer and offline retailer are modeled by using the Nash game theory. The online and offline retailers announce retail quantities q_{d1} and q_{r1}, respectively, and simultaneously at the beginning of the first sales period, then in the promotion sales period, the online retailer and offline retailer set the retail quantity q_{d2} in Model D, or quantity q_{r2} in Model R, or q_{d2} and q_{r2} in Model B.

Correspondingly, the prices of online and offline retailers in the first sales period are p_{d1} and p_{r1}, respectively, and the discount prices in Period 2 are p_{d2} and p_{r2}. Due to the delivery delay in the online channel, consumers may expect that the price in the online channel is less than that in the offline channel to balance the disadvantage, i.e.,

$$p_{d1} \leq p_{r1}, p_{d2} \leq p_{r2}.$$

We refer to the first selling period as the full-price period and the second selling period as the promotion period, that is,
$$p_{d2} \leq p_{d1}, p_{r2} \leq p_{r1}, p_{r2} \leq p_{d1}.$$

We also consider that in this dual-channel supply chain, firms are risk-neutral, and the information between the two channels is symmetric. Finally, we follow the literature (e.g., [24,25]) to normalize the production and selling costs of the products to zero.

3.2. Strategic Consumers

On the one hand, a product is worth v subject to a real inspection while immediate possession has worth θv when the product is obtained from the direct channel due to the delivery delay in the online channel. Here parameter $\theta \in (0,1)$ is called the consumer acceptance of the direct channel. Specifically, when θ approaches to 1, it indicates that the delivery time is very short, and the consumer tends to fully accept the direct channel, and vice versa. On the other hand, patient consumers may decide to buy products in the second period and wait until the price is low enough. To incorporate this behavior, parameter $\rho \in [0,1)$ is used to interpret as a measure of the consumer's patience. As with [8,26], the parameter ρ could also be explained as product availability in the second period because there exists out of stock in the second period. The structure of the consumer utility function here is the same as in the above-mentioned literatures. Since we focus on strategic consumers' behaviors in our paper, we define it as the degree of consumer patience. Throughout the analysis, a consumer is "myopic" when $\rho = 0$ and "infinite" when ρ approaches to 1 [14]. The acceptance of online channel $0 < \theta < 1$ arises from the delay due to the delivery time in the online channel. Consumers are willing to wait more time for the second-period promotional activities to get low prices; thus, for the same products, consumers would obtain a smaller discount when buying from the second period in comparison to buying from the online channel, which means $\theta \geq \rho$ as they are both referring to the delay in consumption.

Strategic consumers are homogenous in the valuation of products, the consumption value (alternatively called "willingness to pay") v is assumed following uniformly distributed on [0,1] within the consumer population. Thus, a consumer evaluates four expected utilities of different strategies as follows.

(1) Buy at the beginning of the first selling period (denoted "Period 1") through the offline retail channel, which yields an expected utility of $u_{r1} = v - p_{r1}$.
(2) Buy at the beginning of Period 1 through the direct online channel, which yields an expected utility of $u_{d1} = \theta v - p_{d1}$.
(3) Wait for the second selling period (denoted "Period 2") through the offline channel, the expected utility is $u_{r2} = \rho(v - p_{r2})$;
(4) Wait for the second selling period (denoted "Period 2") through the online channel, the expected utility is $u_{d2} = \rho(\theta v - p_{d2})$.

Where p_{ij} is the price of the i channel in Period j, where $i = d, r$ and $j = 1, 2$. Here d and r denote the direct online channel and offline retail channel, respectively.

3.3. Demand Functions

We need to analyze consumers' choice when products can be purchased in dual channels in both periods on four common scenarios: Model N, Model D, Model R, and Model B. Figure 2 illustrates the utility functions in the case of these models.

Generally speaking, $\frac{p_{d2}}{\theta}$ (or p_{r2}) is a threshold value that consumers have a positive utility buying through an online channel (or an offline channel) in Period 2. The consumer who is indifferent between buying products through an offline channel and an online retail channel in Period 1 is located at $\frac{p_{r1}-p_{d1}}{1-\theta}$, which could be observed intuitively in Figure 2a. Then, for a consumer who prefers the online channel in Period 1 to the online channel in Period 2. The consumer's valuation v should exceed $\frac{p_{d1}-\rho p_{d2}}{\theta(1-\rho)}$, which is shown in Figure 2b,d. Next, from Figure 2c,d, we can obtain that each consumer whose valuation exceeds $\frac{p_{d1}-\rho p_{r2}}{\theta-\rho}$ would consider buying from the online retailer in Period 1 rather than buying in the offline channel in Period 2.

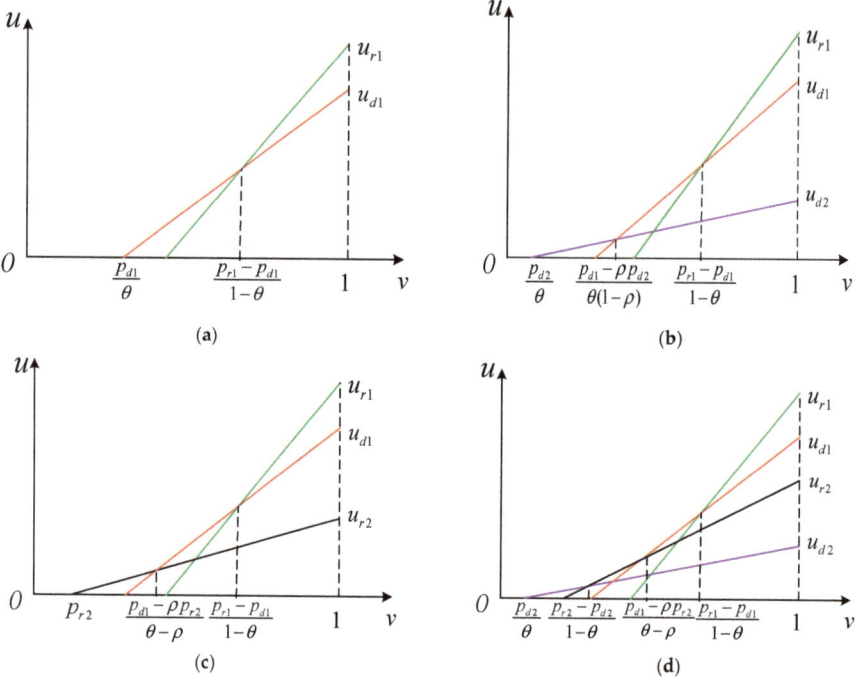

Figure 2. Consumer utilities. (**a**) Model N; (**b**) Model D; (**c**) Model R; (**d**) Model B.

Thus, with the help of consumer utilities and Figure 2, we obtain the demand functions in each model. In Model N, the demand functions of the online and offline retailers are as follows:

$$q_{r1} = 1 - \frac{p_{r1} - p_{d1}}{1-\theta}, \quad q_{d1} = \frac{p_{r1} - p_{d1}}{1-\theta} - \frac{p_{d1}}{\theta}. \tag{1}$$

For the situation of Model D, the demand functions are

$$q_{r1} = 1 - \frac{p_{r1} - p_{d1}}{1-\theta}, \quad q_{d1} = \frac{p_{r1} - p_{d1}}{1-\theta} - \frac{p_{d1} - \rho p_{d2}}{\theta(1-\rho)}, \quad q_{d2} = \frac{p_{d1} - \rho p_{d2}}{\theta(1-\rho)} - \frac{p_{d2}}{\theta}. \tag{2}$$

In the case of Model R, the demand functions are

$$q_{r1} = 1 - \frac{p_{r1} - p_{d1}}{1-\theta}, \quad q_{d1} = \frac{p_{r1} - p_{d1}}{1-\theta} - \frac{p_{d1} - \rho p_{r2}}{\theta - \rho}, \quad q_{r2} = \frac{p_{d1} - \rho p_{r2}}{\theta - \rho} - p_{r2}. \tag{3}$$

Lastly, in Model B, the demand functions are

$$q_{r1} = 1 - \frac{p_{r1} - p_{d1}}{1-\theta}, \quad q_{d1} = \frac{p_{r1} - p_{d1}}{1-\theta} - \frac{p_{d1} - \rho p_{r2}}{\theta - \rho},$$
$$q_{r2} = \frac{p_{d1} - \rho p_{r2}}{\theta - \rho} - \frac{p_{r2} - p_{d2}}{1-\theta}, \quad q_{d2} = \frac{p_{d1} - \rho p_{r2}}{\theta - \rho} - \frac{p_{d2}}{\theta}. \tag{4}$$

4. Model Analysis

In this section, we model the game and analyze the equilibrium outcomes under the four different scenarios (as described in Section 3.1) when the online and offline retailers conduct the Nash game, where firms choose quantities rather than prices, e.g., [3,17,24,27].

4.1. No Promotion Model (Model N)

Here, we discuss the scenario that the second period is not introduced. This scenario is used as a benchmark model with which to compare the three subsequent models. In Model N, the demand function is shown in Equation (1). Hence, the reverse function is

$$p_{r1} = 1 - q_{r1} - \theta q_{d1}, \, p_{d1} = \theta(1 - q_{r1} - q_{d1}).$$

In this benchmark model, the online and offline retailers decide their retail quantities at the same time. The profit functions of online and offline retailers are as follows:

$$\max_{q_{r1}} \pi_r = (1 - q_{r1} - \theta q_{d1})q_{r1}, \max_{q_{d1}} \pi_d = \theta(1 - q_{r1} - q_{d1})q_{d1}.$$

Therefore, the equilibrium results are obtained as the following proposition.

Proposition 1. *In Model N, the online and offline retailer's optimal sales quantities q_{r1}^N and q_{d1}^N are $q_{r1}^N = \frac{2-\theta}{4-\theta}$, $q_{d1}^N = \frac{1}{4-\theta}$. Corresponding, the optimal prices are $p_{r1}^N = \frac{2-\theta}{4-\theta}, p_{d1}^N = \frac{\theta}{4-\theta}$.*

Proof. For given π_r and π_d, calculate the first derivate of on q_{r1} and q_{d1}, respectively. We have $\frac{\partial \pi_r}{\partial q_{r1}} = -2q_{r1} - \theta q_{d1} + 1 = 0$, $\frac{\partial \pi_d}{\partial q_{d1}} = -q_{r1} - 2q_{d1} + 1 = 0$, therefore, there exists a unique optimal pair $q_{r1}^N = \frac{2-\theta}{4-\theta}, q_{d1}^N = \frac{1}{4-\theta}$. Substituting q_{r1}^N and q_{d1}^N into $p_{r1}^N = 1 - q_{r1}^N - \theta q_{d1}^N, p_{d1}^N = \theta(1 - q_{r1}^N - q_{d1}^N)$, the prices in dual channels are $p_{r1}^N = \frac{2-\theta}{4-\theta}, p_{d1}^N = \frac{\theta}{4-\theta}$. □

Arya et al. (2007) studied the same problem as in Model N without considering strategic consumers [3], while in our paper, we examine how acceptance of the online channel affects the whole pricing strategy of dual channels. Proposition 1 elaborates that in Model N, the sale price decided by the online retailer increases with rising acceptance of the online channel. Intuitively, as the competitor of the online retailer, the performance of offline retailer results has declined with the increasing acceptance of the online channel. This observation is partly consistent with the property of the offline retailer in Chiang et al. (2003) [4]. They stated that when acceptance of the online channel is below a cannibalistic threshold, the optimal prices and quantities are not influenced by the acceptance of the online channel. However, we further find that the acceptance of the online channel plays an important role for the online retailer no matter how small of consumer acceptance of the direct channel is.

4.2. Direct Online Channel Introduced in Period 2 (Model D)

This subsection considers Model D, in which the offline retailer does not do any promotions, and their second-period price is the same as the price in Period 1. As a result, their second-period sales volume is zero. In the case of Model D, the online and offline retailers' demand functions are stated by Equation (2). Hence, the reverse demand functions are

$$q_{d1} = \theta(1 - q_{r1} - q_{d1} - \rho q_{d2}), \, p_{r1} = 1 - q_{r1} - \theta q_{d1} - \rho \theta q_{d2}, \\ p_{d2} = \theta(1 - q_{r1} - q_{d1} - q_{d2}). \tag{5}$$

The online retailer and the offline retailer interact as follows: the online retailer determines the optimal sale quantity q_{d1}, and the offline retailer chooses the optimal sale quantity q_{r1} at the same time in Period 1. Then the online retailer decides the sales quantity q_{d2} sold through the direct online channel in Period 2.

We then use backward induction to determine the perfect equilibriums. Proposition 2 shows that there exists a unique Nash equilibrium for this model.

Proposition 2. *There exists a unique Nash equilibrium in the Model D*

$$q_{r1}^D = \frac{8-6\rho-4\theta+2\rho\theta+\rho^2\theta}{2(8-6\rho-2\theta-\rho\theta+2\rho^2\theta)}, q_{d1}^D = \frac{(1-\rho)(2-\rho\theta)}{8-6\rho-2\theta-\rho\theta+2\rho^2\theta}.$$

Further, the corresponding optimal prices in Period 1 and solutions in Period 2 are

$$p_{r1}^D = \frac{(2-\rho\theta)(8-6\rho-4\theta+2\rho\theta+\rho^2\theta)}{4(8-6\rho-2\theta-\rho\theta+2\rho^2\theta)},$$
$$p_{d1}^D = \frac{\theta(2-\rho)(4-2\rho-2\rho\theta+\rho^2\theta)}{4(8-6\rho-2\theta-\rho\theta+2\rho^2\theta)},$$

$$p_{d2}^D = \frac{\theta(4-2\rho-2\rho\theta+\rho^2\theta)}{4(8-6\rho-2\theta-\rho\theta+2\rho^2\theta)}, q_{d2}^D = \frac{4-2\rho-2\rho\theta+\rho^2\theta}{4(8-6\rho-2\theta-\rho\theta+2\rho^2\theta)}.$$

Proof. See Appendix A. □

Zhou (2016) also studied the pricing problem in a dual-channel considering online promotion [19]. They focused on the relationship between selling cost advantages and price strategies chosen by retailers, while we concentrate on the effects of consumers' strategic behaviors on the optimal quantities and prices of retailers. Following the preceding discussion of the optimal decisions in Model D, the following propositions are provided.

Proposition 3. *In Model D* $\frac{\partial p_{d2}^D}{\partial \rho} > 0$, $\frac{\partial q_{d2}^D}{\partial \rho} > 0$, $\frac{\partial q_{r1}^D}{\partial \rho} > 0$, $\frac{\partial q_{d1}^D}{\partial \rho} < 0$ *and* $\frac{\partial q_{r1}^D}{\partial \theta} < 0$, $\frac{\partial p_{d2}^D}{\partial \theta} > 0$, $\frac{\partial q_{d2}^D}{\partial \theta} > 0$, $\frac{\partial q_{d1}^D}{\partial \theta} > 0$.

Proof. See Appendix A. □

Proposition 3 shows that when the acceptance of the online channel is high, or consumers prefer to wait, online retailers strategically raise the price in Period 2 to obtain a high marginal benefit. What is more, as consumers become more patient, the online retailer extends the sales volume in Period 2 and reduces it in Period 1 in response to alleviate the competition pressure with the offline retailer in Period 1. From the offline retailer's viewpoint, offering an online channel in Period 2 is a serious threat to the offline retailer. To compete against the online channel and protect its market share, the offline retailer has to cut its retail price as the only effective tool. Thus, the offline retailer's sales volume increases. Finally, as the direct channel becomes more attractive, more consumers would choose to buy from the online channel instead of the offline channel.

4.3. Retail Offline Channel Introduced in Period 2 (Model R)

In Model R, the online and offline retailers' demand functions are stated by Equations (3). Thus, the reverse functions in Model R are

$$p_{d1} = \theta(1-q_{r1}-q_{d1}) - \rho q_{r2}, \ p_{r1} = 1-q_{r1}-\theta q_{d1}-\rho q_{r2},$$
$$p_{r2} = 1-q_{r1}-q_{d1}-q_{r2}. \tag{6}$$

The behaviors of two retailers in Model R in Period 1 are similar, as in Model D. Proposition 4 shows that there exists a unique Nash equilibrium for Model R.

Proposition 4. *In the Model R, the optimal retail quantities in Period 1 are*

$$q_{r1}^R = \frac{2-\rho-\theta}{4-2\rho-\theta}, q_{d1}^R = \frac{2-\rho}{2(4-2\rho-\theta)}.$$

Further, the corresponding optimal prices in Period 1 and equilibrium solutions in Period 2 are

$$p_{r1}^R = \frac{(2-\rho)(4-\rho-2\theta)}{4(4-2\rho-\theta)}, p_{d1}^R = \frac{(2-\rho)(2\theta-\rho)}{4(4-2\rho-\theta)}, p_{r2}^R = q_{r2}^R = \frac{2-\rho}{4(4-2\rho-\theta)}.$$

Proof. See Appendix A. □

Following the preceding discussion of the optimal decisions in the Model R, the following propositions are provided.

Proposition 5. *In Model R*, $\frac{\partial p_{r2}^R}{\partial \rho} > 0$, $\frac{\partial q_{r2}^R}{\partial \rho} > 0$, $\frac{\partial p_{r1}^R}{\partial \rho} < 0$, $\frac{\partial q_{r1}^R}{\partial \rho} < 0$, $\frac{\partial p_{d1}^R}{\partial \rho} < 0$, $\frac{\partial q_{d1}^R}{\partial \rho} < 0$, and $\frac{\partial p_{d1}^R}{\partial \theta} > 0$, $\frac{\partial q_{d1}^R}{\partial \theta} > 0$, $\frac{\partial p_{r2}^R}{\partial \theta} > 0$, $\frac{\partial q_{r2}^R}{\partial \theta} > 0$, $\frac{\partial p_{r1}^R}{\partial \theta} < 0$, $\frac{\partial q_{r1}^R}{\partial \theta} < 0$.

Proof. See Appendix A. □

In the offline channel, more patient consumers in Period 1 would shift to purchase in Period 2. As a result, the sales volume through the offline channel in Period 1 decreases with the degree of patience, while the sales volume in Period 2 increases. But it seems counter-intuitive that more patient consumers will not wait for lower prices in Period 2. Confronted with patient consumers, offline retailers' marketing strategy of setting a high price in Period 2 aimed at attracting consumers to purchase in Period 1 since they could obtain a high marginal profit in Period 1 since $p_{r1}^R \geq p_{r2}^R$.

Propositions 5 also shows that when the offline retailer extends the selling period, confronted with more patient consumers, the offline retailer sets a lower selling price to avoid reducing fierce competition in Period 1. Then the online retailer has to cut its selling price to protect its market share. Accordingly, the online retailer's sales volume in Period 1 increases with respect to the degree of patience, which is symmetric with properties of offline retailer's sales volume in Model D, as shown in Proposition 3.

Contrary to the existing literature of single sales period, Chiang et al. [4] and Xu et al. [7] verified that offline retailers' price is negatively correlated with the acceptance of online channels. Propositions 5 shows that in Period 2, the offline retailer aggressively increases prices to make the competition more intense with the rising acceptance of the online channel, the main reason is that the offline retailer occupies the entire second-period market in Model R.

4.4. Both Channels Introduced in Period 2 (Model B)

In Model B, the online and offline retailers' demand functions are stated by Equation (4), it follows that the reverse demand functions are

$$p_{d1} = \theta(1 - q_{r1} - q_{d1} - \rho q_{d2}) - \rho q_{r2}, p_{r1} = 1 - q_{r1} - \theta q_{d1} - \rho q_{r2} - \rho\theta q_{d2},$$
$$p_{d2} = \theta(1 - q_{r1} - q_{d1} - q_{r2} - q_{d2}), p_{r2} = 1 - q_{r1} - q_{d1} - q_{r2} - \theta q_{d2}.$$
(7)

We discuss the optimal quantity decision in Model B, which means that the first-period and second-period sales quantities will be announced at the beginning of the corresponding selling period. Using backward induction, Proposition 6 shows that a unique Nash equilibrium exists for Model B.

Proposition 6. *In Model B, the optimal retail quantities in Period 1 are*

$$q_{r1}^B = \frac{(4\theta - \theta^2 - 2\rho)\left[(2-\theta)(4-\theta)^2 - 2\rho(8 - 5\theta + \theta^2)\right] - 2\rho\theta(1-\theta)(4-\theta)}{(4\theta - \theta^2 - 2\rho)\left[(4-\theta)^3 - 2\rho(16 - 6\theta + \theta^2)\right] - 4\rho\theta(1-\theta)(4-\theta)},$$

$$q_{d1}^B = \frac{(4-\theta-2\rho)\left[(4-\theta)(4\theta-\theta^2-2\rho)-2\rho\theta\right]}{(4\theta-\theta^2-2\rho)\left[(4-\theta)^3-2\rho(16-6\theta+\theta^2)\right]-4\rho\theta(1-\theta)(4-\theta)}.$$

Further, equilibrium solutions in Period 2 are

$$q_{r2}^B = \frac{(2-\theta)(4-\theta-2\rho)(4\theta-\theta^2-2\rho)}{(4\theta-\theta^2-2\rho)\left[(4-\theta)^3-2\rho(16-6\theta+\theta^2)\right]-4\rho\theta(1-\theta)(4-\theta)^2},$$

$$q_{d2}^B = \frac{(4-\theta-2\rho)(4\theta-\theta^2-2\rho)}{(4\theta-\theta^2-2\rho)\left[(4-\theta)^3-2\rho(16-6\theta+\theta^2)\right]-4\rho\theta(1-\theta)(4-\theta)^2}.$$

The corresponding optimal prices in Period 1 and Period 2 could be obtained by Equations (7).

Proof. See Appendix A. □

5. Discussion and Comparison

In this section, the effect of strategic consumer behaviors on optimal results in Model B is discussed, and the optimal results obtained under four different decision strategies are compared.

5.1. Sensitivity Discussion

First, take a closer look at the implications of consumer patience for two parties' outcomes in Model B. When the acceptance of the direct channel is given, taking $\theta = 0.8$, the properties of sales volumes and prices about ρ, $\rho \leq 0.8 = \theta$ will be represented in Figures 3 and 4. Taking $\rho = 0.4$, we obtain the properties of sales volumes and prices about θ, $\theta \geq 0.4 = \rho$, which are shown in Figures 5 and 6, respectively.

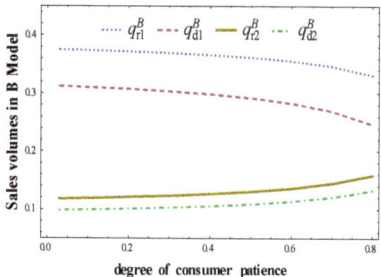

Figure 3. Impact of ρ on sales volumes in dual channels Model B.

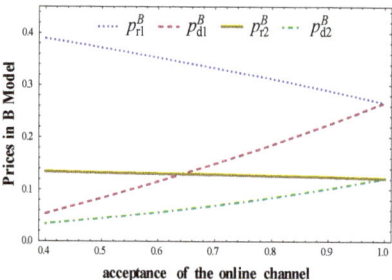

Figure 4. Impact of ρ on prices in Model B.

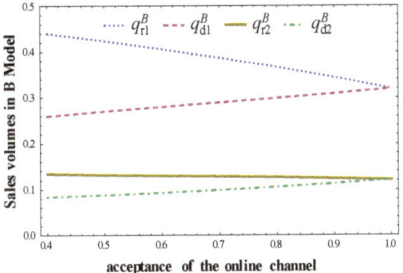

Figure 5. Impact of θ on sales volumes in Model B.

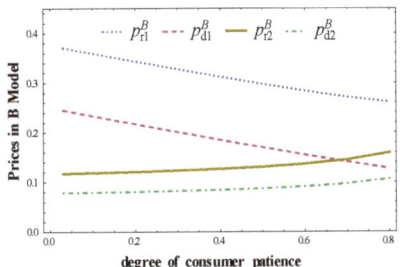

Figure 6. Impact of θ on prices in Model B.

Some discussions about these figures are given in the following:

In Model B, Figures 3 and 4 reveal that sales volumes and prices in Period 1 decrease with the rising degree of patience either from the online or offline retailer. While in Period 2, sales volumes and prices have opposite characters, that is, $\frac{\partial p_{i1}^B}{\partial \rho} < 0, \frac{\partial q_{i1}^B}{\partial \rho} < 0, \frac{\partial p_{i2}^B}{\partial \rho} > 0, \frac{\partial q_{i2}^B}{\partial \rho} > 0, i = d, r$. The reason for this is that retailers aim at attracting consumers to purchase in Period 1 since they have a high margin in Period 1. This relationship seems to contradict to the expected results from consumers, which suggests that more patient consumers could wait for the lower price. In fact, this interesting phenomenon, that second-period prices decrease as consumer's patience increases, is also investigated by Papanastasiou and Savva (2016), where they only considered one retail channel in the two-period model [14].

In addition, the price and sales volumes set by the online retailer increase as the acceptance of the online channel increases in both Period 1 and Period 2. While it is the opposite for all the results of the offline retailer, i.e., $\frac{\partial p_{dj}^B}{\partial \theta} > 0, \frac{\partial q_{dj}^B}{\partial \theta} > 0, \frac{\partial p_{rj}^B}{\partial \theta} < 0, \frac{\partial q_{rj}^B}{\partial \theta} < 0, j = 1, 2$, which is illustrated in Figures 5 and 6. Obviously, this efficient online channel helps increase online prices and forces the offline retailer to markdown retail prices in both selling periods, which is different from the results in Model R. Proposition 5 stated that the offline retailer aggressively increases the offline second-period price as online channels gain acceptance because the offline retailer occupies the entire market.

5.2. Comparison of Dominating Areas

In this section, we graphically show some regions with the pair of parameters (θ, ρ) where we compare selling quantity sequences in Period 1, the online retailer's profit sequences, and offline retailer's profit sequences under four different models. Some discussions about these figures are given in the following.

First, the relationship of offline quantities and online prices with respect to the acceptance of the direct channel will be examined by numerical experiments by taking $\theta = 0.8$ in the left-hand of Figures 7–10. Similarly, we set $\rho = 0.4$ in the right-hand of Figures 7–10 to investigate the offline quantities and online prices regarding the degree of consumer patience. The conclusion could be graphically obtained without any assumption in region $\{(\theta,\rho)|0 \leq \rho \leq \theta \leq 1\}$, but it is not clear in 3-Dimension situation. Thus, we assume $\theta = 0.8$ or $\rho = 0.4$, respectively, to investigate the properties in 2-Dimensional figures. From Figures 7–10, in the region $\{(\theta,\rho)|0 \leq \rho \leq \theta \leq 1\}$, the quantity and price sequences in Period 1 under four models are stated in the following proposition.

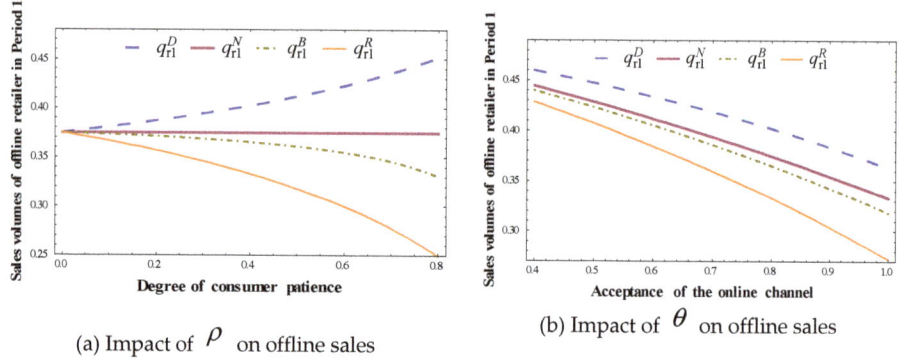

Figure 7. Offline quantities sequences in four sales strategies.

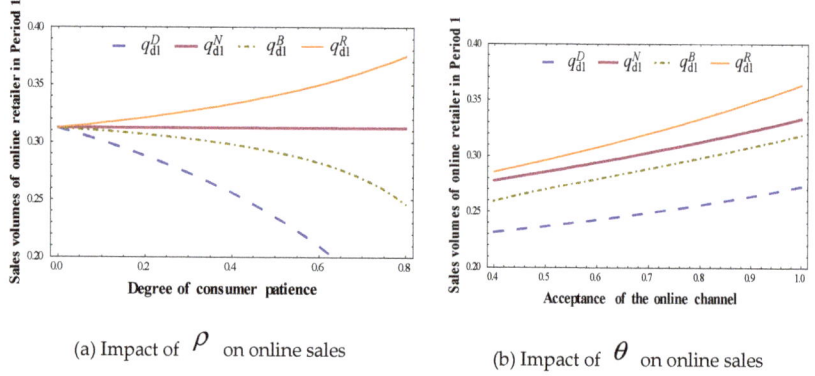

Figure 8. Online quantities sequences in four sales strategies.

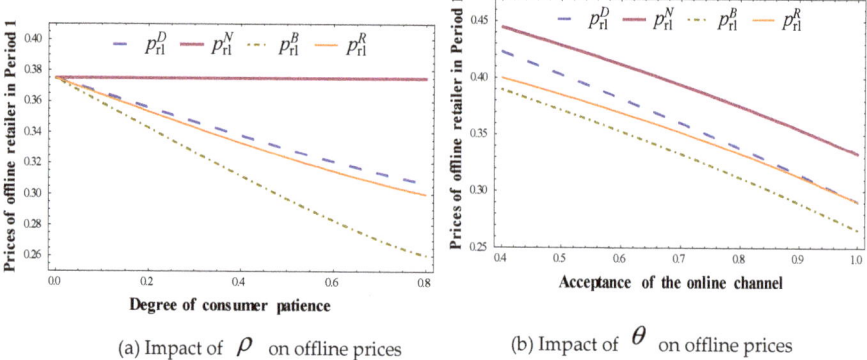

Figure 9. Offline prices sequences in four sales strategies.

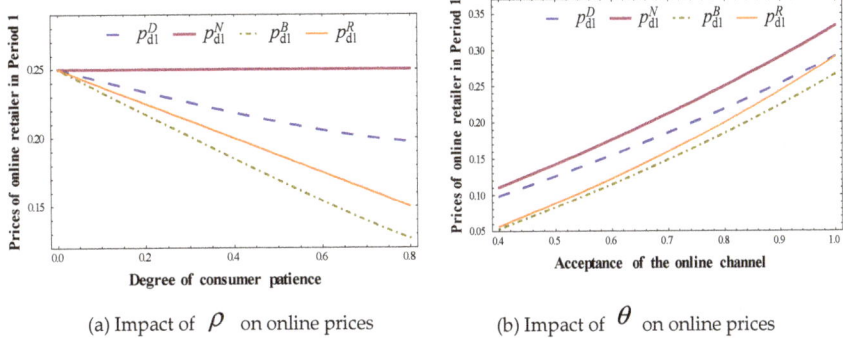

(a) Impact of ρ on online prices   (b) Impact of θ on online prices

Figure 10. Online prices sequences in four sales strategies.

Figures 7–10 verify the monotonic properties of quantities and prices with respect to parameters, as stated in Propositions 3, 5 and 7. The acceptance of the online channel has a positive impact on online retailer's prices and quantities, while it has a negative impact on the offline retailer's prices and quantities. All quantities in Period 1 except for offline retailer's quantity are decreasing with the increase in the degree of consumer patience because of inter-periodic competition. However, counter-intuitively, as shown in the left-hand of Figure 7, the sales volume of the offline retailer increases with the increase in the degree of consumer patience in Model D. In Model D, offering the online channel in Period 2 is a serious threat to the offline retailer. To compete against the online channel and protect its market share in Period 1, the offline retailer has to cut its retail price as the only effective tool even confronted with patient consumers. Thus, the offline retailer's sales volume increases. As displayed in the left-hand of Figure 8, the sales volume of the online retailer also increases with increasing degree of consumer patience in Model R for a similar reason.

Some discussions about quantities or prices sequences in these figures are given in the following:

- $q_{r1}^D \geq q_{r1}^N \geq q_{r1}^B \geq q_{r1}^R$, $q_{d1}^R \geq q_{d1}^N \geq q_{d1}^B \geq q_{d1}^D$. Specifically, the online or offline retailer would achieve higher sales quantities when the second-period promotion sales are provided by its competitor rather than itself.
- $p_{i1}^N \geq p_{i1}^D \geq p_{i1}^R \geq p_{i1}^B$, $i = d, r$. That is, offline and online retailers have the same price sequence. Further, the retail price without promotion strategy is the highest.

It reveals that the first-period quantity of the online retailer is highest when sales promotion is provided by the offline retailer; the offline retailer also has a symmetrical property. This is partly because there is no chance for consumers to purchase through the online channel in Period 2 in Model R. Moreover, sales quantities of both the online retailer and the offline retailer reach their second-highest in Model N and third-highest in Model B. On the other hand, their second-highest prices are obtained in Model D. Channel competition only exists in Model N, the online and offline retailers set the highest prices in Period 1 in Model N, while channel competition and periodic competition both exist in Model B, retailers set the lowest prices in Model B. We can thus conclude that Model N or Model D will be preferable for online and offline retailers as they set high volumes and prices for the first period, which will be verified below.

Similar to Arya et al. [3] and Xiong et al. [17], compared with only the offline channel setting, the manufacturer, also acting as an online retailer, may offer a lower wholesale price to the downstream retailer in the dual-channel setting. Our results also show that, compared with promotion only through the offline channel (i.e., Model R), the online retailer would set a lower first-period price when promoting sales through dual channels (i.e., Model B) to offset the advantage of online retailer's competitive position in the retail market. Among other results, we found that the first-period price was always higher in Model D than that in Model R. Moreover, Erhun et al. [28] stated that the first-period price increases as the number of selling period increases. However, we show that the first-period price in the single-period setting (i.e., Model N) is the highest.

Without any assumptions, we can graphically summarize the detailed profit sequences of the online and offline retailers under four different strategies in the following figures.

From the region $\{(\theta, \rho)|0 \leq \rho \leq \theta \leq 1\}$ shown in Figure 11, the online retailer in Model N dominates in terms of profit. π_d^R is the secondary dominator when θ and ρ take higher values, or π_d^D is the secondary dominator when θ and ρ take lower values. Surprisingly, π_d^B is always the lowest in all regions except for the higher values for θ and ρ, where π_d^D is the minimum. For the offline retailer, π_r^D dominates when ρ and θ is higher, but when ρ takes lower values irrespective of the θ value, π_r^N dominates others. In the region that Model D dominates, π_r^N is the secondary dominator and π_r^R is the minimum. While in the region that Model N dominates, π_r^D is the secondary dominator and π_r^B is the minimum (see Figure 12).

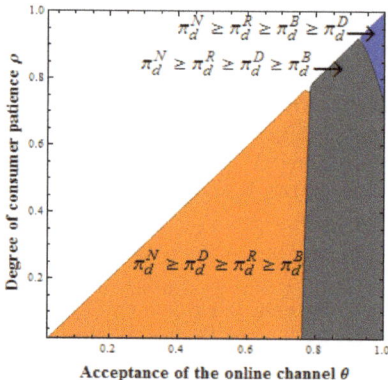

Figure 11. Online retailer's profit sequences under four sales strategies.

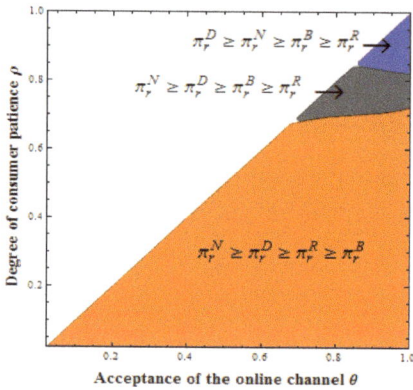

Figure 12. Offline retailer's profit sequences under four sales strategies.

The optimal channel structure strategy depends on the players confronting which types of consumers. In summary, the best choice for the online retailer is Model N, while it is better to choose Model N or Model D than Model R or Model B for the offline retailer. In particular, compared with providing an offline channel in Period 2 by themselves, the offline retailer could get higher profits when the online retailer introduces the online channel in Period 2. The main reason is that the offline retailer's selling quantity and price in Period 1 in Model D are initially higher than those in the Model

R. Practically speaking, offline retailers would be more enthusiastic on Cyber Monday and Double Eleven than even Black Friday.

Contrary to the conventional wisdom, Erhun et al. (2008) stated that all members within the supply chain benefit from multi-period trading [28]. Figures 11 and 12 imply that bearing the brunt of the online channel, single-period selling dominates two-period selling in the dual-channel supply chain. In other words, the profits of online and offline retailers in the single-period dual-channel model (i.e., N Model) is always higher than those in the two-period dual-channel model (i.e., B Model). Thus, offering dual channels in the promotion period will not make the best strategy. We believe this difference stems from the competition between online and offline retailers rather than the competition between the upstream manufacturer and the downstream retailer.

On the other hand, Figure 11 shows that, from the online retailer viewpoint, compared with only the offline channel existing in the second period (i.e., Model R), the introduction of the online channel in the second sales period (i.e., Model B) cannot create a higher profit. Figure 12 also shows that the offline channel achieves a higher profit in Model D rather than in Model B. Our observation differs from the observation of Chiang et al. [4] and Chen [29] who argued that "introducing the online channel model provides the online retailer's profit improvement". They focus on determining the optimal selling price in a single-period supply chain while we consider the pricing policy in a two-period supply chain.

6. Discussion and Comparison

Dual-channel distribution systems, including an online channel and an offline channel, have been adopted by many retailers (defined as "Model N"). Retailers consider setting the promotion period to sell products through only the online channel, or only the offline channel, or dual channels at the same time (defined as "Model D", "Model R", and "Model B", respectively). Although the marketing issues associated with two selling periods have been well studied, limited results are known about how these marketing strategies are influenced by the retailers' adoption of different channel structures and the penetration of strategic consumers. To fill in this gap, this paper attempts to investigate how strategic consumers' behaviors impact the optimal quantities and how the retailers' selling strategies affect their profits.

We first introduced four original demand functions of the dual-channel supply chain by incorporating the consumers' utility in our paper. Based on it, we establish optimal quantity and price strategies for online/offline retailers. Our results indicate that, in most cases, all online retailer's optimal sales volumes and prices increase as the acceptance of the online channel increases, while it is opposite for the offline retailer. But in Model R, the offline retailer's optimal decisions in the second period increase with the increase in the acceptance of the online channel. In addition, opposite to all results in Period 1, all volumes and prices obtained in Period 2 increase with the degree of patience increasing. It seems counter-intuitive that a more patient consumer would not wait for a lower price in Period 2.

At last, the equilibrium outcomes of both parties are analyzed among four different channel structure strategies by graphical analysis. We found that strategic consumers' behaviors (i.e., the acceptance of the online channel and the degree of patience) could strongly influence the choice of channel structures for online and offline retailers. It reveals that online and offline retailers earn the highest profit in the single-period model except when consumers have a high degree of patience and a high acceptance of the online channel. Finally, the offline retailer could obtain higher profits when the online retailer introduces an online channel in the promotion period instead of providing an offline channel in the promotion period.

In this paper, only the retailers and consumers are considered, while future studies could be extended to consider a manufacturer or supplier to analyze the wholesale price. In addition, consumers are homogenous in this paper. Studies next will consider different types of consumers, for example, experienced and inexperienced consumers, which is closer to real life. Moreover, dynamic modeling would seem to be more appropriate in two-period framework research, so dynamic investigation would be our subsequent research topic.

Author Contributions: Conceptualization, Q.L. and J.H.; writing—original draft preparation, Q.L.; writing—review and editing, Q.L. and F.H.; project administration, J.H. All authors have read and agreed to the published version of the manuscript.

Funding: Natural Science Foundation of China through Grant Number 71873111, 71802100, 71273214 and Humanities and Social Science Project of Education Committee through Grant Number 18YJAZH024.

Conflicts of Interest: The authors declare no conflict of interest.

Appendix A

Proof of Proposition 2. In Period 2, the online retailer's problem is to choose the optimal quantity q_{d2} to maximize profit

$$\max_{q_{d2}} \pi_{d2} = p_{d2}q_{d2} = \theta(1 - q_{r1} - q_{d1} - q_{d2})q_{d2}$$

with q_{r1} and q_{d1} given. Thereafter, the online retailer's optimal retail quantity

$$q_{d2} = (1 - q_{r1} - q_{d1})/2. \tag{A1}$$

By substituting Equation (A1) into Equation (5), based on it, the profit functions of the online retailer and offline retailer could be expressed as

$$\max_{q_{r1}} \pi_{r1} = p_{r1}q_{r1} = \frac{1}{2}[(2 - \rho\theta) - (2 - \rho\theta)q_{r1} - \theta(2 - \rho)q_{d1}]q_{r1},$$

$$\max_{q_{d1}} \pi_{d1} = p_{d1}q_{d1} + \rho p_{d2}q_{d2} = \frac{\theta(2 - \rho)}{2}(1 - q_{r1} - q_{d1})q_{d1} + \frac{\rho\theta}{4}(1 - q_{r1} - q_{d1})^2.$$

From the online retailer's profit, the first term is the revenue from selling in Period 1, and the second term is revenue from Period 2. The profit in Period 2 is multiplied by the degree of patience, which is mainly due to the patience of the consumer working as a discount factor, i.e., the online retailer discounts the profits in Period 2 to get the profit value at the same time point as the value in the Period 1. The first-order optimality conditions are

$$\begin{cases} \frac{\partial(2\pi_{r1})}{\partial q_{r1}} = -2(2 - \rho\theta)q_{r1} - \theta(2 - \rho)q_{d1} + (2 - \rho\theta) = 0, \\ \frac{\partial(2\pi_{d1})}{\partial q_{d1}} = -2(1 - \rho)q_{r1} - (4 - 3\rho)q_{d1} + 2(1 - \rho) = 0. \end{cases}$$

Solving the above equations, we get

$$q_{r1}^D = \frac{8 - 6\rho - 4\theta + 2\rho\theta + \rho^2\theta}{2(8 - 6\rho - 2\theta - \rho\theta + 2\rho^2\theta)}, \quad q_{d1}^D = \frac{(1 - \rho)(2 - \rho\theta)}{8 - 6\rho - 2\theta - \rho\theta + 2\rho^2\theta}.$$

Substituting in Equations (A1) and (5), we obtain p_{r1}^D, p_{d1}^D, p_{d2}^D, q_{d2}^D as stated in Proposition 3. □

Proof of Proposition 3. In Model D, for convenience, we define $A = \frac{4 - 2\rho - 2\rho\theta + \rho^2\theta}{8 - 6\rho - 2\theta - \rho\theta + 2\rho^2\theta}$, $\alpha = 8 - 6\rho - 2\theta - \rho\theta + 2\rho^2\theta$. Taking the derivative of A with respect to ρ, we have

$$\alpha^2 \frac{\partial A}{\partial \rho} = 2(1 - \theta)(4 - 2\rho - 2\rho\theta + \rho^2\theta)$$
$$+ 2\theta(1 - \rho)(2\rho + 2\theta - 3\rho\theta) + \theta[4\rho(1 - \theta) + \rho\theta^2(2\theta - \rho)] > 0.$$

Combined with $q_{d2}^D = \frac{A}{4}, p_{d2}^D = \frac{\theta}{4}A$, we can derive $\frac{\partial q_{d2}^D}{\partial \rho} > 0, \frac{\partial p_{d2}^D}{\partial \rho} > 0$. Then, taking the derivative of q_{r1}^D and q_{d1}^D with respect to ρ,

$$2\alpha^2 \frac{\partial q_{r1}^D}{\partial \rho} = \theta\left[(2-\rho)(2-2\theta+2\rho\theta-\rho^2\theta) + (1-\rho)[2(1-\theta)+(1-\rho)(6-2\theta)+\rho^2\theta]\right] > 0,$$

$$\alpha^2 \frac{\partial q_{d1}^D}{\partial \rho} = (2-\rho\theta)(-2+2\theta-8\rho+\rho^2\theta) + \theta(1-\rho)(-6-4\rho+2\theta-\rho^2\theta) < 0,$$

we find $\frac{\partial q_{r1}^D}{\partial \rho} > 0, \frac{\partial q_{d1}^D}{\partial \rho} > 0$. At last, we consider the properties of sales quantities regarding θ. By first partial derivatives, we obtain

$$4\alpha^2 \frac{\partial q_{d2}^D}{\partial \theta} = 2(1-\rho)(2-\rho)^2 > 0,$$
$$4\alpha^2 \frac{\partial p_{d2}^D}{\partial \theta} = \rho(2-\rho)(2+\rho-2\rho^2) + 4(2-\rho)(4-3\rho)(1-\rho\theta) > 0,$$
$$2\alpha^2 \frac{\partial q_{r1}^D}{\partial \theta} = -(8-6\rho)(1-\rho)(2-\rho) < 0,$$
$$\alpha^2 \frac{\partial q_{d1}^D}{\partial \theta} = (1-\rho)(4-4\rho+2\rho^2) > 0.$$

□

Proof of Proposition 4. In Period 2, the offline retailer decides the sale quantity q_{r2} sold through the offline channel in Period 2. Using backward induction, the offline retailer's profit in Period 2 is

$$\max_{q_{r2}} \pi_{r2} = p_{r2}q_{r2} = (1 - q_{r1} - q_{d1} - q_{r2})q_{r2}.$$

Thus, the optimal retail quantity is

$$q_{r2} = (1 - q_{r1} - q_{d1})/2. \quad (A2)$$

Combining (A2) with Equations (6), we could obtain the optimal prices p_{r1} and p_{d1} in Period 1. Therefore, the profits of online and offline retailers are

$$\max_{q_{r1}} \pi_{r1} = p_{r1}q_{r1} + \rho p_{r2}q_{r2} = \frac{1}{2}[2 - \rho - (2-\rho)q_{r1} - (2\theta - \rho)q_{d1}]q_{r1} + \frac{\rho}{4}(1 - q_{r1} - q_{d1})^2$$

$$\max_{q_{d1}} \pi_{d1} = p_{d1}q_{d1} = \frac{2\theta - \rho}{2}(1 - q_{r1} - q_{d1})q_{d1}.$$

The first-order optimality condition is

$$\frac{\partial(2\pi_{r1})}{\partial q_{r1}} = -(4-3\rho)q_{r1} - 2(\theta - \rho)q_{d1} + 2(1-\rho) = 0,$$
$$\frac{\partial(2\pi_{d1})}{(2\theta - \rho)\partial q_{d1}} = -q_{r1} - 2q_{d1} + 1 = 0,$$

which leads to $q_{r1}^R = \frac{2-\rho-\theta}{4-2\rho-\theta}, q_{d1}^R = \frac{2-\rho}{2(4-2\rho-\theta)}$. Substituting in Equations (A2) and (6), we obtain $p_{r1}^R, p_{d1}^R, p_{r2}^R, p_{d2}^R$ as stated in Proposition 4. □

Proof of Proposition 5. In Model R, define $\beta = 4 - 2\rho - \theta$, taking the derivative of all results obtained in proposition 4 with respect to ρ, we learn

$$2\beta^2 \frac{\partial q_{d1}^R}{\partial \rho} = \theta > 0, \beta^2 \frac{\partial q_{r1}^R}{\partial \rho} = -\theta < 0,$$

$$2\beta^2 \frac{\partial p_{r1}^R}{\partial \rho} = -(1-\theta)(4-2\rho-\theta) - (\theta-\rho)(2-\rho) < 0,$$

$$4\beta^2 \frac{\partial p_{r2}^R}{\partial \rho} = 4\beta^2 \frac{\partial q_{r2}^R}{\partial \rho} = \theta > 0,$$

$$4\beta^2 \frac{\partial p_{d1}^R}{\partial \rho} = -\left[8(1-\rho) - 2(\theta - \rho^2) - 2\theta(\theta - \rho)\right].$$

Combined with $1 - \rho > \theta - \rho > \theta - \rho^2$, $\frac{\partial p_{d1}^R}{\partial \rho} < 0$ could be obtained. The other conclusions can be derived by first-order partial derivative easily, so we omit them here. □

Proof of Proposition 6. Using backward induction, the profits of online and offline retailers can be expressed as

$$\begin{cases} \max_{q_{d2}} \pi_{d2} = p_{d2} q_{d2} = \theta(1 - q_{r1} - q_{d1} - q_{r2} - q_{d2}) q_{d2}, \\ \max_{q_{r2}} \pi_{r2} = p_{r2} q_{r2} = (1 - q_{r1} - q_{d1} - q_{r2} - \theta q_{d2}) q_{r2}. \end{cases}$$

It follows that the optimal solutions can be described as follows

$$q_{d2} = \frac{1 - q_{r1} - q_{d1}}{4 - \theta}, q_{r2} = \frac{(2-\theta)(1 - q_{r1} - q_{d1})}{4 - \theta}. \tag{A3}$$

We rewrite the profits of two retailers for $\pi_{d1} = p_{d1} q_{d1} + \rho p_{d2} q_{d2}$ and $\pi_{r1} = p_{r1} q_{r1} + \rho p_{r2} q_{r2}$ as follows

$$\begin{cases} \max_{q_{d1}} \pi_{d1} = \frac{4\theta - \theta^2 - 2\rho}{4-\theta}(1 - q_{r1} - q_{d1})q_{d1} + \frac{\rho\theta}{(4-\theta)^2}(1 - q_{r1} - q_{d1})^2, \\ \max_{q_{r1}} \pi_{r1} = \left[1 - \frac{2\rho}{4-\theta} - (1 - \frac{2\rho}{4-\theta})q_{r1} - (\theta - \frac{2\rho}{4-\theta})q_{d1}\right]q_{r1} + \frac{\rho(2-\theta)^2}{(4-\theta)^2}(1 - q_{r1} - q_{d1})^2. \end{cases}$$

The derivative of the profit function with respect to the selling quantity is

$$\begin{cases} -2[\gamma - \rho(2-\theta)^2]q_{r1} - [\gamma - 2\rho(2-\theta)^2]q_{d1} + \gamma + (1-\theta)(4-\theta)^2 - 2\rho(2-\theta)^2 = 0, \\ -(\gamma - 2\rho\theta)q_{r1} - 2(\gamma - \rho\theta)q_{d1} + \gamma - 2\rho\theta = 0, \end{cases}$$

where $\gamma = (4-\theta)(4\theta - \theta^2 - 2\rho)$, and consequently, we get q_{r1}^B and q_{d1}^B as expressed in Proposition 6. Substituting q_{r1}^B and q_{d1}^B in (A3) and (7), yields the other optimal results. □

References

1. Hardman, D.; Simon, H.; Ashok, N. *Keeping Inventory-And Profits-Off the Discount Rack*; White Paper Booz Allen Hamilton Inc.: McLean, VA, USA, 2007.
2. Ji, G.J.; Zhao, Y. Strategic consumer behavior analysis based on dual channels and sale periods. *Int. J. u-e-Serv. Sci. Technol.* **2015**, *8*, 135–152. [CrossRef]
3. Arya, A.; Mittendorf, B.; Sappington, D.E. The bright side of supplier encroachment. *Mark. Sci.* **2007**, *26*, 651–659. [CrossRef]
4. Chiang, W.Y.K.; Chhajed, D.; Hess, J.D. Direct marketing, indirect profits: A strategic analysis of dual-channel supply-chain design. *Manag. Sci.* **2003**, *49*, 1–20. [CrossRef]
5. Li, G.; Li, L.; Sethi, S.P.; Guan, X. Return strategy and pricing in a dual-channel supply chain. *Int. J. Prod. Econ.* **2019**, *215*, 153–164. [CrossRef]
6. Hua, G.; Wang, S.; Cheng, T.E. Price and lead time decisions in dual-channel supply chains. *Eur. J. Oper. Res.* **2010**, *205*, 113–126. [CrossRef]

7. Xu, H.; Liu, Z.Z.; Zhang, S.H. A strategic analysis of dual-channel supply chain design with price and delivery lead time considerations. *Int. J. Prod. Econ.* **2012**, *139*, 654–663. [CrossRef]
8. Chen, K.Y.; Kaya, M.; Özer, Ö. Dual sales channel management with service competition. *Manuf. Serv. Oper. Manag.* **2008**, *10*, 654–675. [CrossRef]
9. Zhang, R.; Li, J.; Huang, Z.S.; Liu, B. Return strategies and online product customization in a dual-channel supply chain. *Sustainability* **2019**, *11*, 3482. [CrossRef]
10. De Giovanni, P.; Zaccour, G. A two-period game of a closed-loop supply chain. *Eur. J. Oper. Res.* **2014**, *232*, 22–40. [CrossRef]
11. Lin, Z. Price promotion with reference price effects in supply chain. *Transp. Res. Part E-Logist. Transp. Rev.* **2016**, *85*, 52–68. [CrossRef]
12. Maiti, T.; Giri, B.C. Two-period pricing and decision strategies in a two-echelon supply chain under price-dependent demand. *Appl. Math. Model.* **2017**, *42*, 655–674. [CrossRef]
13. Chen, K.; Xiao, T. Ordering policy and coordination of a supply chain with two-period demand uncertainty. *Eur. J. Oper. Res.* **2011**, *215*, 347–357. [CrossRef]
14. Papanastasiou, Y.; Savva, N. Dynamic pricing in the presence of social learning and strategic consumers. *Manag. Sci.* **2016**, *63*, 919–939. [CrossRef]
15. Lai, G.; Debo, L.G.; Sycara, K. Buy now and match later: Impact of posterior price matching on profit with strategic consumers. *Manuf. Serv. Oper. Manag.* **2010**, *12*, 33–55. [CrossRef]
16. Huang, S.; Yang, C.; Zhang, X. Pricing and production decisions in dual-channel supply chains with demand disruptions. *Comput. Ind. Eng.* **2012**, *62*, 70–83. [CrossRef]
17. Xiong, Y.; Yan, W.; Fernandes, K.; Xiong, Z.K.; Guo, N. "Bricks vs. Clicks": The impact of manufacturer encroachment with a dealer leasing and selling of durable goods. *Eur. J. Oper. Res.* **2012**, *217*, 75–83. [CrossRef]
18. Yan, W.; Li, Y.; Wu, Y.; Palmer, M. A rising e-channel tide lifts all boats? The impact of manufacturer multichannel encroachment on traditional selling and leasing. *Discret. Dyn. Nat. Soc.* **2016**, *2016*, 2898021. [CrossRef]
19. Hübner, A.H.; Kuhn, H.; Sternbeck, M.G. Demand and supply chain planning in grocery retail: An operations planning framework. *Int. J. Retail Distrib. Manag.* **2013**, *41*, 512–530. [CrossRef]
20. Cachon, G.P.; Swinney, R. Purchasing, pricing, and quick response in the presence of strategic consumers. *Manag. Sci.* **2009**, *55*, 497–511. [CrossRef]
21. Giampietri, E.; Verneau, F.; Del Giudice, T.; Carfora, V.; Finco, A. A Theory of Planned behaviour perspective for investigating the role of trust in consumer purchasing decision related to short food supply chains. *Food. Qual. Prefer.* **2018**, *64*, 160–166. [CrossRef]
22. Zimon, D.; Pedro, D. Proposal of a concept for improving the sustainable management of supply chains in the textile industry. *Fibers Text. East. Eur.* **2018**, *26*, 8–12. [CrossRef]
23. Zhou, C. Pricing model for dual sales channel with promotion effect consideration. *Math. Probl. Eng.* **2016**, *2016*, 1804031. [CrossRef]
24. Li, H.; Leng, K.; Qing, Q.; Zhu, S.X. Strategic interplay between store brand introduction and online direct channel introduction. *Transp. Res. Part E-Logist. Transp. Rev.* **2018**, *118*, 272–290. [CrossRef]
25. Li, G.; Li, L.; Sun, J. Pricing and service effort strategy in a dual-channel supply chain with showrooming effect. *Transp. Res. Part E-Logist. Transp. Rev.* **2019**, *126*, 32–48. [CrossRef]
26. Liu, Q.; Van Ryzin, G.J. Strategic capacity rationing to induce early purchases. *Manag. Sci.* **2008**, *54*, 1115–1131. [CrossRef]
27. Li, H.; Shao, J.; Zhu, S.X. Parallel importation in a supply Chain: The impact of gray market structure. *Transp. Res. Part E-Logist. Transp. Rev.* **2018**, *114*, 220–241. [CrossRef]
28. Erhun, F.; Keskinocak, P.; Tayur, S. Dynamic procurement, quantity discounts, and supply chain efficiency. *Prod. Oper. Manag.* **2008**, *17*, 543–550. [CrossRef]
29. Chen, T. Effects of the pricing and cooperative advertising policies in a two-echelon dual-channel supply chain. *Comput. Ind. Eng.* **2015**, *87*, 250–259. [CrossRef]

© 2019 by the authors. Licensee MDPI, Basel, Switzerland. This article is an open access article distributed under the terms and conditions of the Creative Commons Attribution (CC BY) license (http://creativecommons.org/licenses/by/4.0/).

Article

Intelligence in Tourism Management: A Hybrid FOA-BP Method on Daily Tourism Demand Forecasting with Web Search Data

Keqing Li [1], Wenxing Lu [1,2,*], Changyong Liang [1,2] and Binyou Wang [1]

1. School of Management, Hefei University of Technology, Hefei 230009, China; keqinglhfut@gmail.com (K.L.); cyliang@hfut.edu.cn (C.L.); wbylrxzyy@163.com (B.W.)
2. Ministry of Education Key Laboratory of Process Optimization and Intelligent Decision-making, Hefei University of Technology, Hefei 230009, China
* Correspondence: luwenxing@hfut.edu.cn; Tel.: +86-180-5604-4985

Received: 17 May 2019; Accepted: 5 June 2019; Published: 11 June 2019

Abstract: The Chinese tourism industry has been developing rapidly for the past several years, and the number of people traveling has been increasing year by year. However, many problems still beset current tourism management. Lack of effective management has caused numerous problems, such as tourists stranded during tourist season and the declining service quality of scenic spots, which have become the focus of tourists' attention. Network search data can intuitively reflect the attention of most users through the combination of the network search index and the back propagation (BP) neural network model. This study predicts the daily tourism demand in the Huangshan scenic spot in China. The filtered keyword in the Baidu index is added to the hybrid neural network, and a BP neural network model optimized by a fruit fly optimization algorithm (FOA) based on the web search data is established in this study. Different forecasting methods are compared in this paper; the results prove that compared with other prediction models, higher accuracy can be obtained when it comes to the peak season using the FOA-BP method that includes web search data, which is a sustainable means of practically solving the tourism management problem by a more accurate prediction of tourism demand of scenic spots.

Keywords: tourism management; hybrid method; fruit fly optimization algorithm; neural network; web search data; forecast of daily tourism demand; optimization method

1. Introduction

The Chinese tourism industry has grown along with development of the Chinese economy. According to statistics, the number of inbound and domestic tourists in China are increasing year by year and the tourism industry is developing rapidly [1]. Such rapid growth has resulted in tourism management problems requiring urgent solutions, including forecasting tourism demand especially when large numbers of tourists travel to scenic areas for short-term visits. Management is under considerable pressure, and any negligence can cause serious public safety problems. For instance, on 2 October 2013, many tourists were stuck at the entrance of Jiuzhaigou Valley because of overcrowding. To prevent this from happening again and ensure tourism develops healthily and sustainably, the forecast of tourist flow, especially short-term forecasting, is an important research direction.

Forecasting methods used in the past include an econometric model [2], a time series model [3,4], artificial neural network and support vector machine, and hybrid methods [5]. Most of these techniques are based on historical data, but the long lag period of the predicted values often leads to problems. At present, an increasing number of scholars regard network search as an important and leading source of research data and timely information [6]. When people search for information on the Internet, their search record can reflect their concerns.

This study aims to propose an effective short-term tourism demand forecasting method that can effectively forecast daily tourist flow on the basis of web search data and back propagation (BP) neural network optimized by a fruit fly optimization algorithm (FOA). Meanwhile, the hybrid neural network contains selected web search data in order to optimize the prediction effect of the model, which is proved to be effective in this study.

This paper proceeds as follows: Section 2 gives the background of this study. Section 3 proposes the process by which the model is built to forecast the tourist flow using network search data. Section 4 provides the process of empirical study. Section 5 presents the result and evaluation of our study. Finally, Section 6 discusses the contributions of this work and implications for further research.

2. Background

2.1. Research on Tourism Demand Forecasting

In recent years, a number of research on the prediction of tourist flow and various prediction methods have been proposed, including an econometric model, a time series model, artificial neural network and support vector machine, and hybrid methods [5]. Econometric models [7,8], which are widely used for forecasting tourism demand, analyze the causal relationships between dependent variables (tourism demand) and explanatory variables (influencing factors). A time series model is always based on historical data, and many scholars use this model to research tourism demand forecasting [9]. With the development of computer technology, artificial intelligence methods such as artificial neural network and support vector machine are being applied in the research on tourism demand forecasting [10,11]. Some scholars propose hybrid methods by combining an econometric model and artificial intelligence method or time series models [12]. No single model works in every situation because different models have their own applicable occasions [5]. Genetic algorithm is used to optimize neural network to improve prediction accuracy [13]. A vector error correction model (VECM) was used for forecasting the tourism demand of Jeju Island [14]. Neural networks (NN) are used to improve the accuracy of the Grey–Markov (GM) forecasting model [15]. An autoregressive integrated moving average (ARIMA) model is used to predict the urban residents' future travel rate in five years [16].

The researches of the above scholars focus on the tourism demand forecasting in a long-term forecasting such as several years and months. Daily forecasting is rarely studied by scholars, and deserves further study, and which is concerned in this study.

2.2. Forecasting with Network Search Data

Internet search is an important channel for people to search for information. With the popularity of mobile Internet, people can more conveniently inquire information through search engines [6]. In recent years, the use of network data has also provided a new data source and analysis basis for social science. Ginsberg et al. [17] used the Google search engine to determine whether the online search index of flu-related keywords was highly correlated with the number of people with influenza in the same period. The researchers successfully predicted the trend of influenza outbreak and proved that the search index has a certain predictive ability for the epidemic. The method of network search index as a prediction tool quickly spread in different research directions. Valuable research achievements that have used this method include predictions on a film box office [18], consumer confidence index [19], unemployment [20], and stock market [21].

Baidu is the largest search engine in China and has the most users. Research also shows that when studying consumer behavior, Baidu search has higher predictive power than Google search [22].

3. Methodology

This study proposes a complete modeling process—from selecting the keywords, getting the Baidu index, analyzing the correlation between daily Baidu index and actual total tourist flow, and choosing the index with the most correlation, to building forecasting models and evaluating the corresponding performance. The steps are shown as follows:

(1) Select the keywords. This step depends on the process by a Chinese traveler who is considering the plan to visit a scenic spot. Before traveling, he/she may search for information about destination, weather, strategy, price, hotel, and the like on the Internet. The traveler defines the keywords related to his/her destination.

(2) Obtain the corresponding keywords from Baidu index for correlation analysis. The Baidu index is based on the search volume of netizens in Baidu. It takes keywords as the statistical object to scientifically analyze and calculate the weighted sum of the search frequency of each keyword in Baidu web search.

(3) Considering the lag in web search, set the lag time and analyze the correlation between the Baidu index and actual total tourist flow. The most relevant lag period will be selected.

(4) The improved FOA algorithm is used to optimize the BP neural network. A hybrid FOA-BP model is established in this study in order to predict daily tourism demand.

(5) Determine the parameters in the model in order to train the model.

(6) Evaluate the accuracy. Genetic algorithms–back propagation (GA-BP) neural network and particle swarm optimization–back propagation (PSO-BP) neural network are selected as the benchmark models. The mean absolute percentage error (MAPE) is selected as the evaluation standard.

Current research on keyword selection method has not reached a consensus yet. At present, the three main methods of keyword selection [6] are as follows: technical method, direct method, and scope method. With the technical method, all possible keywords should be brought into the research scope via high-performance computer technology. The direct method determines the keyword directly by subjective experience. The scope method initially determines the range of a choice of words and then selects keywords within the range.

3.1. Back Propagation Neural Network

The neural network of error BP training algorithm, or BP network, is a multilayer feedforward network with hidden layers, which systematically solve the problem of learning the connection weight of hidden units in multilayer networks. If the number of input nodes of the network is M and the number of output nodes is L, then the neural network can be regarded as the mapping from M-dimensional Euclidean space to L-dimensional Euclidean space. This mapping is highly nonlinear. The structure diagram of the BP network is shown in Figure 1.

Its basic principle is the gradient maximum drop method, and the central idea is to adjust the weight (W_{ij}, W_{ki}) to minimize the total error of the network. Gradient search technology is adopted to minimize the error mean square of the actual output value and the expected output value of the network. In the network learning process, the error (e_k) propagates back and corrects the weight coefficient. The vulnerability to many external factors, such as weather, weekend, holidays, and so on, causes tourist traffic to be highly nonlinear. Traditional statistical models cannot easily show these complex nonlinear features [23]. Owing to the strong nonlinear mapping ability of the BP neural network, it is selected as the prediction model of this study.

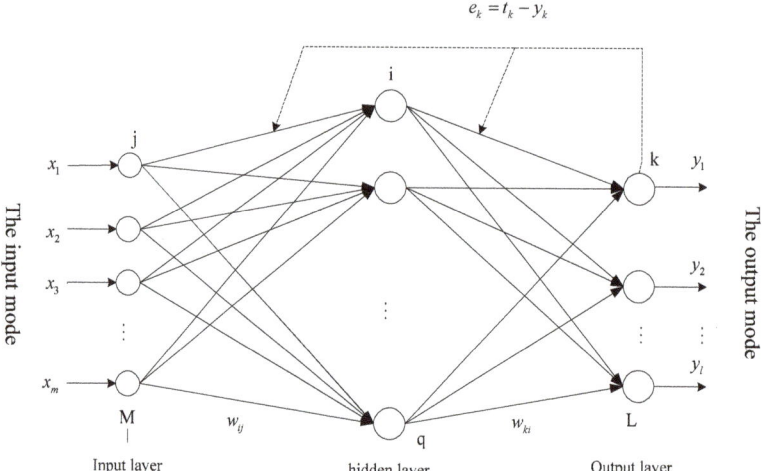

Figure 1. Back propagation neural network.

3.2. Fruit Fly Optimization Algorithm

There are several intelligent algorithms before such as genetic algorithms (GA) proposed in 1975 by Holland [24], and the particle swarm optimization (PSO) proposed in 1995 by Eberhart [25]. Fruit fly optimization algorithm (FOA) is a new swarm intelligent algorithm proposed in 2011 by Wen-Tsao Pan [26], and at present the research is still in the initial stage [27]. The algorithm is simple in process, with few control parameters and easy to implement. Some scholars have successfully applied it to structural engineering design optimization problems [28], wireless sensor network layout [29], resource-constrained project scheduling problems [30] and other fields.

Standard fruit fly optimization algorithm steps are as follows:

Step 1: Initialize population size, termination criterion, and the fruit fly swarm location (a_o, b_o).

Step 2: Foraging with smell. Individual flies use their sense of smell to find random distances and directions for food. Generate the ith location of fruit fly randomly as (a_i, b_i):

$$\begin{cases} a_i = a_o + RandomValue \\ b_i = b_o + RandomValue \end{cases} \tag{1}$$

where *RandomValue* means random distances and directions.

Step 3: First, the distance from the origin (Dist) is estimated, and then the determination value of smell concentration (S) is calculated, which is the reciprocal of the distance:

$$Dist_i = \sqrt{(a_i^2 + b_i^2)} \tag{2}$$

$$S_i = \frac{1}{Dist_i} \tag{3}$$

Step 4: Substitute S_i into the smell concentration determination function to find the smell concentration:

$$Smell_i = Function(S_i) \tag{4}$$

Step 5: Find the fruit flies with the lowest concentration of smell (strive for the minimum):

$$bestSmell = \min(Smell_i) \tag{5}$$

Step 6: The fruit fly swarms fly towards this position using vision, and record the best smell concentration and position:

$$\begin{cases} i^* = \operatorname{argmin}\{i : \operatorname{Function}(S_i) = \operatorname{bestSmell} \\ a_0 = a_{i^*} \\ b_0 = b_{i^*} \end{cases} \quad (6)$$

Step 7: Iterate to see if the new Smell is better than the previous one, terminate the search progress when reaching the termination criterion. Otherwise, repeat the steps 2–6 above.

The fruit fly optimization algorithm is shown in Figure 2:

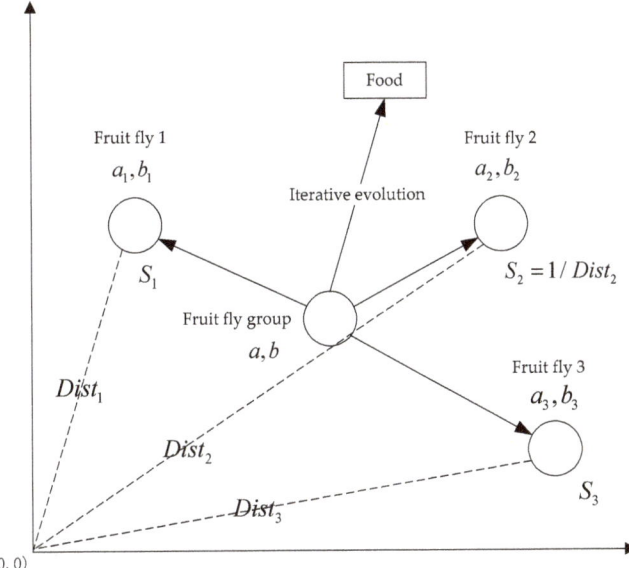

Figure 2. Fruit fly optimization algorithm.

3.3. The Hybrid Fruit Fly Optimization Algorithm-Back Propagation Model with Web Search Data

The hybrid FOA-BP model optimizes the connection weight and threshold of the neural network to improve the generalization ability and learning performance of the BP network, so as to improve the overall search efficiency of the initial BP neural network.

The optimization process is actually the change of fruit flies' position. Meanwhile the learning process of BP neural network is actually the updating process of weights and thresholds. Therefore, it can be considered that the smell concentration value of each fruit fly group corresponds to the weight value and threshold value in the BP network iteration process, the number of fruit fly population depends on the number of parameters to be optimized and the neural network output error of training samples is used as the smell concentration determination function. In the foraging process, the position change of the fruit fly can minimize the error of the network. At the end of each iteration, the fruit fly with the best smell concentration is regarded as the current globally optimal fruit fly. When the training process is repeated repeatedly until the error meeting the requirements or the number of pre-set iterations is reached, the search terminates. At this time, the set of weights and thresholds obtained are the final results.

The whole process in this study is shown in Figure 3 as follows:

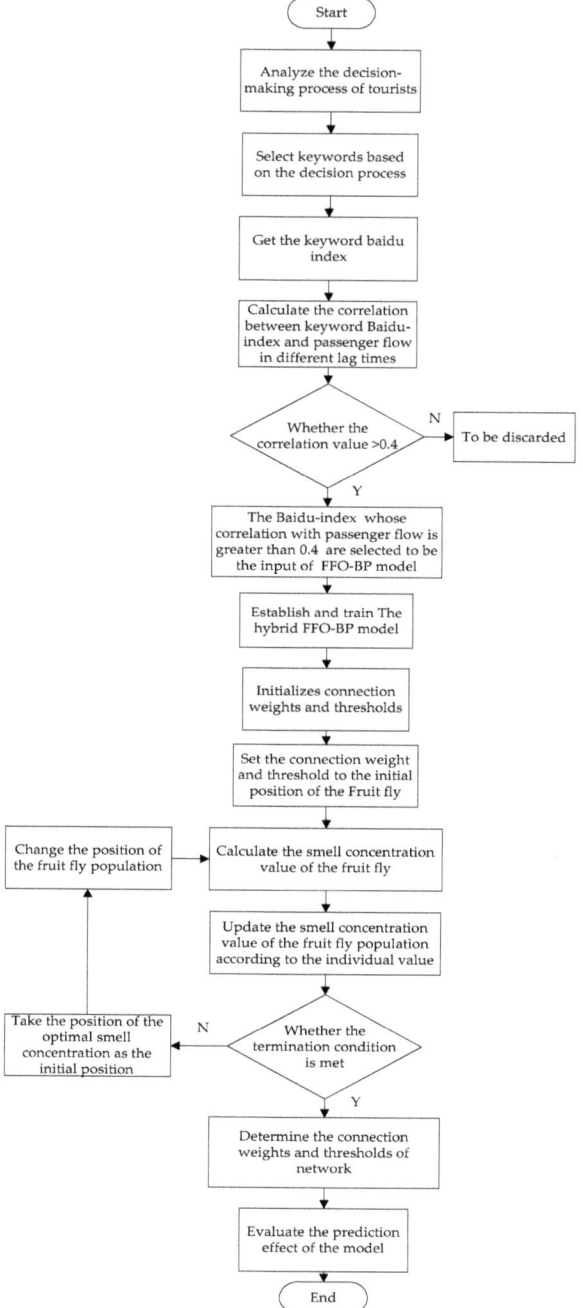

Figure 3. The flow chart.

4. Empirical Study

4.1. Data

This paper takes the Mount Huangshan scenic area as an example of a famous Chinese scenic spot. Mount Huangshan is listed as a United Nations Educational Scientific and Cultural Organization (UNESCO) world natural and cultural heritage site in 1990, and was selected as one of the first global geoparks in 2004. In 2017, there were 3.3687 million visitors from around the world.

The daily historical data selected are from 2015 to 2017, and all data were obtained from the research project in cooperation with the Huangshan scenic area and the network search index (i.e., Baidu index) from the large data in the Baidu search engine. Baidu's massive Internet behavior data are based on a data-sharing platform. The search index is based on the search volume of netizens in Baidu, and it takes keywords as the statistical object to scientifically analyze and calculate the weighted sum of the search frequency of each keyword in the Baidu search engine.

4.2. Keyword Selection

This study comprehensively used the direct method and scope method based on the tourism decision-making process and the related information that visitors would want to focus on before traveling to their destination. This study selected relevant keywords, including "destination", "destination + guide", "destination + travel guides", "destination + tickets", "destination + weather", based on the tourism destination, strategy, ticket price, scenic spots, weather, and accommodation, among many other factors. Finally, 50 initial keywords related to the decision-making process were selected. On the basis of the Huangshan scenic area, this study chose "Huangshan", "Huangshan tourism guide", "Huangshan tickets", "Huangshan guide", "Huangshan weather", and "Huangshan accommodation" from 50 initial keywords as benchmark keywords according to its correlation with passenger flow and queries the corresponding Baidu index. At the same time, a keyword-mining tool (http://tool.chinaz.com/) was employed to verify that the above six keywords are the top keywords. To verify the correlation between keywords and actual number of tourist flow, this study analyzed the correlation between six keywords from Baidu index and the actual number of total daily tourist flow to the Huangshan scenic spot. Table 1 shows that the correlation analysis results with a confidence level of 0.01.

Table 1. Results of correlation analysis.

Keywords	Correlation
Huangshan	0.471
Huangshan tourism guide	0.551
Huangshan weather	0.102
Huangshan guide	0.417
Huangshan accommodation	0.323
Huangshan tickets	0.449

The table reveals that the four keywords with high correlation degrees were "Huangshan", "Huangshan tourism guide", "Huangshan ticket", and "Huangshan tourism guide", with correlations of 0.471, 0.551, 0.449, and 0.417, respectively. Meanwhile, the correlation between the Baidu index of "Huangshan weather" and "Huangshan accommodation" and the total passenger flow was relatively low (less than 0.4). The four keywords with high correlations from the Baidu index were selected as input variables of the prediction model.

4.3. Data Preprocessing

The FOA-BP neural network model in this study was established using MATLAB R2016a software. To improve the prediction accuracy and stabilize the data before using the BP neural network for training prediction, the original data sequence of the total population was normalized to [0, 1] by mapminmax function. The formula is as follows:

$$Y_t = (\frac{X_t - X_{min}}{X_{max} - X_{min}}) \qquad (7)$$

where X_t is the passenger flow on day t in the original one-year data series, X_{min} and X_{max} are the minimum and maximum values of the original sequence, respectively.

4.4. Selection of Input Variables

Data sets in this paper include actual traffic data of the Huangshan scenic spot from 2015 to 2017. To contain the characteristics of the whole data set, data from 2015–2016 were selected as the training set and the 2017 actual tourist flow data were selected as the test set. The tourist flow prediction model was built based on the FOA-BP neural network containing web search data. The past total number of people, weather, weekends, and official holidays were selected as input variables [23], and the Baidu index of relevant keywords was used as the input variable. Weather, weekend, and official holidays were added into the model as dummy variables.

(1) Daily total number of tourists in the past.

Past total number of tourists had four corresponding rules: by date, by week, by total number of tourists last week, and by total number of tourists the week before last. The correlation analysis results of past total number of tourists and target total number of tourists are shown in Table 2.

Table 2. Results of correlation analysis.

Corresponding Rules	By Date	By Week	By Total Number of Tourists Last Week	By Total Number of Tourists the Week before Last Week
Correlation	0.416	0.376	0.258	0.171

Note: "By date" means the rule is that "2017-01-01" corresponds to "2016-01-01".

Therefore, the past total number of tourists corresponded by date is selected as the input variable X_1.

(2) Weather.

Weather was added into the model in the form of the dummy variable X_2:

$$X_2 = \begin{cases} 1 \\ 0 \end{cases}$$

, 1 represents severe weather such as blizzard, heavy snow, moderate snow, heavy rain, heavy rain, thundershowers, and showers; 0 represents non-severe weather such as sunny, cloudy, and drizzle.

(3) Weekend.

Weekend was added into the model in the form of the dummy variable X_3:

$$X_3 = \begin{cases} 1 \\ 0 \end{cases}$$

, 1 represents weekend; 0 represents workday.

(4) Official holiday

Official holiday was added into the model in the form of the dummy variable X_4:

$$X_4 = \begin{cases} 1 \\ 0 \end{cases}$$

, 1 represents official holiday; 0 represents ordinary day.

(5) Baidu index of keywords

Through the previous analysis, we selected four keywords which have the highest correlation: "Huangshan", "Huangshan tourism guide", "Huangshan ticket", and "Huangshan tourism strategy."

Given the lag period between searching information on the Internet and going to travel, this study respectively analyzed the correlation between the Baidu index of the four keywords with a lag period of one day, two days, and three days to a week, and the actual total number of tourists. As shown in Table 3, the Baidu index of keyword with a lag period of two days has the highest correlation with the actual total number of tourists, and was thus selected as the input variable of the model.

Table 3. Results of the correlation analysis of different lag periods.

Lag Period	Correlation with the Total Number of People			
	Huangshan	Huangshan Tourism Guide	Huangshan Guide	Huangshan Tickets
1	0.511	0.629	0.554	0.544
2	0.518	0.638	0.606	0.560
3	0.459	0.581	0.567	0.498
4	0.404	0.512	0.496	0.432
5	0.371	0.458	0.411	0.379
6	0.309	0.408	0.351	0.304
7	0.257	0.372	0.291	0.253

4.5. Building the Model

The tourist flow in peak season (from April to October) presents strong nonlinear characteristics [23]. Compared with the traditional time series prediction model, BP neural network can deal with these complex nonlinear relations well. Usually, the three-layer structure of the BP neural network is enough to reflect the complex nonlinear relationship. Thus, the current study set up a three-layer structure of the FOA-BP neural network. Nine hidden layer nodes and one output layer node were chosen based on many experiments. Previous experiments show that using a sigmoid function as the activation function between the hidden layer and the output layer can have the good prediction effect. As a type of elastic algorithm, trainrp has the advantage of fast convergence speed and small footprint compared with functions such as trainlm, trainscg, and traingd. The trainrp function can likewise achieve better prediction effect, and was selected as the training function of the model in this study.

5. Empirical Results and Evaluation

The GA-BP model, and PSO-BP model were selected as the benchmark model in this study. The prediction result of the model mentioned above were compared with the benchmark model. The input variables of the model we proposed included the past total number of tourists, weather, weekend, official holiday, and Baidu index of keywords (from 0 to 4 keywords). The input variables of the benchmark model only included weather, weekend, and official holiday. In evaluating the prediction effect of the model, MAPE indicator was selected. The formula is as follows:

$$MAPE = \frac{1}{n}(\sum \left|\frac{Y - X_t}{Y}\right|) \times 100, \ t \in 1, 2 \ldots \ldots n \tag{8}$$

where Y represents the actual number, and X_t represents the predicted value.

The comparative experiments in this study were divided into two categories: the prediction results of different models including/excluding the web search data.

Table 4 shows the accuracy of different models excluding the web search data.

Tables 5–7 show the accuracy of different models including the web search data.

According to the predicted results in the tables above, GA-BP, PSO-BP and FOA-BP models including the web search data had a better accuracy than the benchmark models excluding the web search data; the average accuracy of the whole year was improved to a certain extent. Also, more keywords make the result more accurate. From the tables above it can be seen that the average accuracy from April to October was more accurate than in the whole year, especially in June, July and August. Meanwhile, there were also some limitations; the accuracy in January, February, March, November and December was not good enough, but the results were still better than the benchmark models. One of the reasons could be that the actual value is small; usually there are only 1000 to 2500 tourists, which likely leads to a high numerical deviation.

Three models including the web search data have an approximate accuracy for the whole year, however, from the result shown in tables above, the accuracy from April to October of the FOA-BP model is better than that of the GA-BP and PSO-BP models.

The time from April to October is considered the peak season of the Huangshan scenic spot traditionally [23]. The accuracy of the predicted value in this period has more significance for the tourism management of the scenic spot because there are many more tourists in this period. The predicted result and the actual value are shown in Figure 4. Compared with January, February, March, November and December, there are more tourists in this period. As is shown in this study, the FOA-BP model including the web search data achieved the better predicted value compared to benchmark models. As a result, more advice can be provided for the management of scenic spots, such as increasing the corresponding staff to ensure the work efficiency of scenic spots to avoid the occurrence of stranded tourist events.

Table 4. Results of different models excluding the web search data.

Month	MAPE (%)		
	GA-BP	PSO-BP	FOA-BP
1	53.63	58.62	63.41
2	50.99	52.98	49.49
3	46.60	45.09	42.64
4	27.63	32.51	37.52
5	43.63	43.11	36.43
6	32.01	35.99	32.61
7	22.73	20.69	19.23
8	16.83	22.18	18.29
9	31.61	26.93	29.31
10	30.34	28.98	26.10
11	39.00	33.54	36.30
12	57.18	59.43	61.07
The average from April to October	29.23	29.93	28.58
The average for the whole year	37.57	38.25	37.61

Table 5. Results of GA-BP model including the web search data with 1–4 keywords as input.

GA-BP	MAPE (%)			
	One Keyword	Two Keywords	Three Keywords	Four Keywords
1	58.74	45.78	53.94	53.06
2	40.46	33.93	36.54	47.37
3	41.24	41.04	40.12	36.74
4	32.96	26.47	27.64	30.75
5	38.63	35.89	34.30	40.13
6	30.41	30.88	28.06	32.49
7	23.02	23.59	20.68	18.67
8	16.34	17.39	15.49	14.85
9	41.06	34.07	37.21	28.74
10	34.75	28.04	31.45	29.67
11	28.55	25.30	24.92	26.98
12	51.54	32.45	44.47	44.59
The average from April to October	30.91	28.08	27.85	27.84
The average for the whole year	36.36	31.19	32.81	33.47

Table 6. Results of PSO-BP model including the web search data with 1–4 keywords as input.

PSO-BP	MAPE (%)			
	One Keyword	Two Keywords	Three Keywords	Four Keywords
1	54.29	58.31	63.13	43.36
2	40.67	39.47	37.68	50.33
3	46.53	39.31	37.86	37.54
4	31.58	32.96	31.35	30.35
5	36.50	35.75	36.64	34.44
6	35.22	30.59	24.43	25.51
7	18.98	19.16	13.91	15.95
8	26.94	16.74	24.56	20.42
9	34.88	33.52	41.86	36.81
10	30.32	29.89	27.84	26.36
11	30.10	43.27	30.74	31.76
12	43.96	46.77	55.70	48.98
The average from April to October	31.44	28.21	28.71	27.06
The average for the whole year	36.32	35.35	35.46	33.39

Table 7. Results of FOA-BP model including the web search data with 1–4 keywords as input.

FOA-BP	MAPE (%)			
	One Keyword	Two Keywords	Three Keywords	Four Keywords
1	58.66	65.00	65.62	72.91
2	42.04	40.59	39.05	36.40
3	42.12	36.50	36.16	36.72
4	34.03	29.35	30.43	29.57
5	41.84	32.51	33.67	33.90
6	30.78	23.98	22.72	24.59
7	19.58	16.24	13.50	11.80
8	16.32	19.39	16.26	16.42
9	32.04	34.78	39.36	35.71
10	28.48	29.65	29.64	27.49
11	29.11	26.93	28.30	26.00
12	58.25	53.15	52.34	52.14
The average from April to October	28.90	26.45	26.48	25.60
The average for the whole year	36.03	33.94	33.81	33.55

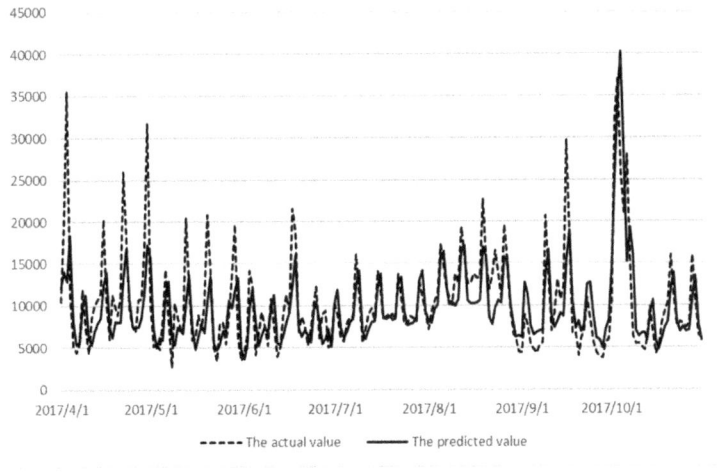

Figure 4. The actual and predicted value.

In general, the hybrid FOA-BP method on daily tourism demand forecasting with web search data proposed in this study can effectively improve the tourist flow prediction accuracy in peak season. Furthermore, the hybrid method proposed was better than other benchmark models with regard to the peak season, which proves the validity of the method in short-term daily tourism demand forecasting.

6. Conclusions and Implications

The rapid development of tourism in recent years has become an important part of the Chinese economy. Thus, the problem of tourism management has become more and more pressing. The prediction of tourist flow, especially short-term tourist flow during peak season, is crucial for tourism management departments. The management department needs to effectively predict future tourism demand, so as to maintain the sustainable development of the scenic spot and avoid damage caused by excessive tourists. A hybrid FOA-BP model is established in this study, which is proved to obtain a more accurate prediction compared with other intelligent algorithms when used in short-term daily tourism demand forecasting. The hybrid model can effectively help the management department to carry on the sustainable management to the scenic spot. Furthermore, taking the famous tourist destination Huangshan scenic spot as an example, this study discusses the application of the Internet search index in the forecast of short-term tourist flow. Moreover, it establishes a model combined with the Baidu index to predict short-term tourist flow. In the selection of Internet search keywords, the benchmark keywords are selected according to the characteristics of the research objects. The benchmark keywords should be reasonable, operable and as comprehensive and accurate as possible. Combining the direct method and the scope method, this study selects the keywords that are related to the destination tourist attractions and have a high search volume, and from which the keywords with a high correlation degree are selected. Considering the lag period between online search and travel, the network search index of the lag period with the highest correlation between related keywords and total number of people is selected through the correlation degree analysis. Experimental results show that the network search index can greatly improve the prediction effect of the original model and is more effective than the benchmark model. However, there are some limitations in the experiment which deserve further study, such as a more accurate keyword selection method and other application methods of web search data in tourism demand forecasting. Generally speaking, the hybrid FOA-BP method proposed in this paper provides a new view for short-term tourism demand forecasting. The proposed method has good prospects in research and application for tourism management, from which the tourism industry can be healthy and sustainable.

Author Contributions: Data curation, K.L.; formal analysis, K.L. and W.L.; methodology, K.L.; supervision, W.L., C.L. and B.W.; writing—original draft, K.L.

Funding: This research received no external funding.

Acknowledgments: This work was supported by the National Natural Science Foundation of China (NSFC) (71331002, 71771075, 71771077, 71601061) and supported by "the Fundamental Research Funds for the Central Universities" (PA2019GDQT0005).

Conflicts of Interest: The authors declare no conflict of interest.

References

1. Bureau, A.S. *Anhui Statistical Yearbook*; China Statistics Press: Beijing, China, 2017.
2. Smeral, E.; Witt, S.F.; Witt, C.A. Econometric forecasts: Tourism trends to 2000. *Ann. Tour. Res.* **1992**, *19*, 450–466. [CrossRef]
3. Rostan, P.; Rostan, A. The versatility of spectrum analysis for forecasting financial time series. *J. Forecast.* **2017**, *37*, 327–339. [CrossRef]
4. Shang, H.L. Forecasting intraday S&P 500 index returns: A functional time series approach. *J. Forecast.* **2017**, *36*, 741–755.

5. Song, H.; Li, G. Tourism demand modelling and forecasting—A review of recent research. *Tour. Manag.* **2008**, *29*, 203–220. [CrossRef]
6. Sun, Y.; Lv, B. A Review of Researches on the Correlation between Internet Search and Economic Behavior. *Manag. Rev.* **2011**, *23*, 72–77.
7. Wong, K.K.F.; Song, H.; Chon, K.S. Bayesian models for tourism demand forecasting. *Tour. Manag.* **2006**, *27*, 773–780. [CrossRef]
8. Song, H.Y.; Witt, S.F. Forecasting international tourist flows to Macau. *Tour. Manag.* **2006**, *27*, 214–224. [CrossRef]
9. Gil-Alana, L.A.; Cunado, J.; Gracia, F.P.D. Tourism in the Canary Islands: Forecasting using several seasonal time series models. *J. Forecast.* **2008**, *27*, 621–636. [CrossRef]
10. Palmer, A.; Montaño, J.J.; Sesé, A. Designing an artificial neural network for forecasting tourism time series. *Tour. Manag.* **2006**, *27*, 781–790. [CrossRef]
11. Hong, W.C. *The Application of Support Vector Machines to Forecast Tourist Arrivals in Barbados: An Empirical Study*; Social Science Electronic Publishing: Rochester, NY, USA, 2006; Volume 23.
12. Song, G.F.; Liang, C.Y.; Liang, Y.; Zhao, S.P.; Management, S.O. Prediction for Tourist Attractions Daily Traffic Based on Modified Genetic Algorithm Optimized BP Neural Network. *J. Chin. Comput. Syst.* **2014**, *20*, 232–238.
13. Huang, H.-C.; Hou, C.I. Tourism Demand Forecasting Model Using Neural Network. International. *J. Comput. Sci. Inf. Technol.* **2017**, *9*, 19–29. [CrossRef]
14. Song, W.L.; Lee, H. The Study on the Tourism Demand Characteristics and Forecasting of Jeju Island. *Tour. Res.* **2018**, *43*, 111–124. [CrossRef]
15. Hu, Y.-C.; Jiang, P.; Lee, P.-C. Forecasting tourism demand by incorporating neural networks into Grey–Markov models. *J. Oper. Res. Soc.* **2018**, *70*, 12–20. [CrossRef]
16. Zong, C.-L.; Wang, L. Prediction of urban residents' travel rate in China based on ARIMA models. *J. Interdiscip. Math.* **2018**, *21*, 1285–1290. [CrossRef]
17. Ginsberg, J.; Mohebbi, M.H.; Patel, R.S.; Brammer, L.; Smolinski, M.S.; Brilliant, L. Detecting influenza epidemics using search engine query data. *Nature* **2009**, *457*, 1012. [CrossRef] [PubMed]
18. Wang, L.; Jia, J.M. Forecasting box office performance based on online search:Evidence from Chinese movie industry. *Syst. Eng.-Theory Pract.* **2014**, *34*, 3079–3090.
19. Vosen, S.; Schmidt, T. Forecasting private consumption: Survey-based indicators vs. Google trends. *J. Forecast.* **2011**, *30*, 565–578. [CrossRef]
20. Francesco, D.A. Predicting unemployment in short samples with internet job search query data. *MPRA Pap.* **2009**, *6*, 1–17.
21. Dimpfl, T.; Jank, S. Can Internet Search Queries Help to Predict Stock Market Volatility? *Eur. Financ. Manag.* **2016**, *22*, 171–192. [CrossRef]
22. Xin, Y.; Pan, B.; Evans, J.A.; Lv, B. Forecasting Chinese tourist volume with search engine data. *Tour. Manag.* **2015**, *46*, 386–397.
23. Liang, C.Y.; Ma, Y.C.; Chen, R.; Liang, Y. The Daily Forecasting Tourism Demand Based on SVR-ARMA Combination Model. *J. Ind. Eng. Eng. Manag.* **2015**, *29*, 122–127.
24. Niazi, M.A. In Memoriam: John Henry Holland—A pioneer of complex adaptive systems research. *Complex Adapt. Syst. Model.* **2015**, *3*, 1–2.
25. Shi, Y.; Eberhart, R.C. Empirical study of particle swarm optimization. In Proceedings of the 1999 Congress on Evolutionary Computation-CEC99 (Cat. No. 99TH8406), Washington, DC, USA, 6–9 July 1999; Volume 1943, pp. 1945–1950.
26. Pan, W.T. *Fruit Fly Optimization Algorithm*; Tsang Hai Book Publishing: Taipei, China, 2011.
27. Lin, S.; Dong, C.; Chen, M.; Zhang, F.; Chen, J. Summary of new group intelligent optimization algorithms. *Comput. Eng. Appl.* **2018**, *54*, 1–9.
28. Du, T.S.; Ke, X.T.; Liao, J.G.; Shen, Y.J. DSLC-FOA: Improved fruit fly optimization algorithm for application to structural engineering design optimization problems. *Appl. Math. Model.* **2018**, *55*, 314–339. [CrossRef]

29. Wu, L. Research on Wireless Sensor Network Layout Based on Improved Fruit Fly Optimization Algorithm. *Microelectron. Comput.* **2016**, *33*, 152–156.
30. Wang, L.; Zheng, X.L. A knowledge-guided multi-objective fruit fly optimization algorithm for the multi-skill resource constrained project scheduling problem. *Swarm Evolut. Comput.* **2018**, *38*, 54–63. [CrossRef]

© 2019 by the authors. Licensee MDPI, Basel, Switzerland. This article is an open access article distributed under the terms and conditions of the Creative Commons Attribution (CC BY) license (http://creativecommons.org/licenses/by/4.0/).

Article

A Bradley-Terry Model-Based Approach to Prioritize the Balance Scorecard Driving Factors: The Case Study of a Financial Software Factory

Vicente Rodríguez Montequín *, Joaquín Manuel Villanueva Balsera, Marina Díaz Piloñeta and César Álvarez Pérez

Department of Project Engineering, University of Oviedo, C/Independencia 3, 33004 Oviedo, Spain
* Correspondence: montequi@api.uniovi.es; Tel.: +34-985-104-272

Received: 16 January 2020; Accepted: 19 February 2020; Published: 19 February 2020

Abstract: The prioritization of factors has been widely studied applying different methods from the domain of the multiple-criteria decision-making, such as for example the Analytic Hierarchy Process method (AHP) based on decision-makers' pairwise comparisons. Most of these methods are subjected to a complex analysis. The Bradley-Terry model is a probability model for paired evaluations. Although this model is usually known for its application to calculating probabilities, it can be also extended for ranking factors based on pairwise comparison. This application is much less used; however, this work shows that it can provide advantages, such as greater simplicity than traditional multiple-criteria decision methods in some contexts. This work presents a method for ranking the perspectives and indicators of a balance scorecard when the opinion of several decision-makers needs to be combined. The data come from an elicitation process, accounting for the number of times a factor is preferred to others by the decision-makers in a pairwise comparisons. No preference scale is used; the process just indicates the winner of the comparison. Then, the priority weights are derived from the Bradley-Terry model. The method is applied in a Financial Software Factory for demonstration and validation. The results are compared against the application of the AHP method for the same data, concluding that despite the simplifications made with the new approach, the results are very similar. The study contributes to the multiple-criteria decision-making domain by building an integrated framework, which can be used as a tool for scorecard prioritization.

Keywords: balanced scorecard; Bradley-Terry; performance evaluation; software factory; multiple-criteria decision-making; AHP

---

1. Introduction

The Balance Scorecard (BSC) framework is probably the most widespread tool to control and manage an organization. The initial proposal was introduced in 1992 by Kaplan and Norton [1,2]. The BSC provides through their perspectives and key performance indicators (KPIs) insights into corporate performance. The BSC provides a template that must be personalized according to the characteristics of each organization. Many authors suggested a set of modifications to customize the initial proposal of the BSC for specific kind of companies or areas.

The BSC framework does not establish the relative importance of its perspectives and indicators, which is a key factor when making decisions and planning strategies. Nevertheless, it can be determined by means of the integration with some multiple-criteria decision-making (MCDM) methods. The prioritizing process of the BSC has been addressed usually following the Saaty's Analytic Hierarchy Process method (AHP) [3], which is one of the most widely used MCDM methods. The relative importance of each criteria is calculated through making pairwise comparisons using a nine-point scale. Several examples can be found in the literature describing use cases. The works from

Clinton et al. [4] and Reisinger et al. [5] were some of the first. Particularly, case studies applying AHP to the BSC have been extensively published. An overview of applications can be consulted in the work of Vaidya and Kumar [6].

When determining the priorities of a BSC, the AHP method is usually applied in a group decision context, collecting the opinions from several decision-makers and determining the preferences of the group as a whole. The opinion of the decision-makers is gathered through an elicitation process by means of pairwise questionnaires. Under this situation, each decision-maker fills a questionnaire with the comparison of each element, and the results are aggregated, usually by means of geometric mean, to arrive at a final solution. Then, the process of AHP is applied for the calculation of the consensual priorities.

Although the Bradley-Terry model is a method that can be used to prioritize criteria, it has been used very little for this purpose, and even less in the context of BSC prioritization. The method is known mainly for the calculation of probabilities in sports tournaments; extensive literature exists regarding this application. Our proposal is to derive the weights of the indicators from the calculation of the Bradley-Terry model, considering that the degree of importance of each indicator will be given by the number of times that each decision-maker has preferred it over the other indicators. Therefore, our method assumes that the prioritization is being carried out in order to get a consensus from a group of decision-makers, and the method is limited to this situation. This is a novel approach that has not been described until now. The Bradley-Terry model is used in this study to determine weights for the perspectives and KPIs included in the BSC. The goal of this paper is to describe the application to this case study and establish a framework that could be replicated in similar scenarios. The results are compared for validation with those provided by the application of the AHP to the same cases.

There are some known issues with the application of the AHP method recognized by academics and practitioners. On the one hand, the number of pairwise comparisons can be very high: $n*(n-1)/2$ for n alternatives/criteria. This could yield that comparisons may be entered in a short amount of time by the decision-makers. On the other hand, there are also concerns about the judgment scale. In Saaty's AHP, the verbal statements are converted into integers from one to nine: the so-called Saaty's fundamental scale. Even though the scale has its own psychophysical basis, as Saaty wrote [7], it is sometimes difficult for the decision-makers to discern between the different intensity levels and, even more, use the same criteria all the time for all the pairwise comparisons. On the other hand, a matter in question is the difficulty of dealing with inconsistent comparisons in the analysis (the decision-maker's arbitrary judgment can lead to some inconsistency). For comparison matrixes that fail the consistency test, the decision-maker has to redo the ratios. To expect the decision-maker to provide the comparisons such that the ranges include only consistent comparison ratios is laborious and highly unrealistic [8]. When the AHP is used to get a group consensus, as is the case of scoring the weights of the BSC indicators by a group of decision-makers, the former issues are emphasized. Then, the chance to get inconsistences is stressed. Under these circumstances, our method can be a very interesting alternative to the utilization of AHP.

The paper discusses to what extent the Bradley-Terry model can simplify the calculation of priorities or avoid those issues. Bradley-Terry does not entail a reduction of the needed pairwise comparisons, but the comparisons are simpler because it does not use a scale of intensity. The Bradley-Terry model only needs the winner option for the calculation, which is noticed as a win-to-loss scale. As consequence, the level of inconsistency is also reduced. In addition, Bradley-Terry accept missing comparisons; that is, when one of the decision-makers is not clear about one of the comparisons, there is no need to fill it out, and the calculations can still be performed.

The remainder of this paper is organized as follows. First, a literature review is introduced. Next, the materials and methods are included. The context where the study was conducted is described, as well as a general description of the BSC. A short introduction to the AHP method is also included alongside a discussion about the points that could be simplified using the Bradley-Terry model. Then, the Bradley-Terry model and its integration with BSC are presented, and the research method is stated. Finally, the results are shown and discussed, and the conclusions and practical implications are exposed.

2. Literature Review

Applications of the Bradley-Terry model are many and varied. Traditionally, sport has been one of the most prominent areas from the beginning. In fact, the model is usually described in a context of sport tournaments, where a set of teams or players confront each other. There are well-documented examples of the use of the model for baseball [9], tennis [10], or basketball [11]. Additional examples can be drawn from the work of Király and Qian [12]. The model has been also used for ranking scientific journals [13], market research [14], or social analysis [15], among others. Another field of application is psychometric, where comparisons are made by different human subjects between pairs of items in terms of preferences [16]. That is the approach followed in our method: the decision-makers express their preferences over the different criteria and the KPIs included in the BSC. A general review about Bradley-Terry applications can be found in the work of Cattelan [17].

The basis for deriving priorities with Bradley-Terry has been introduced in the beginning by Dykstra [18] and more recently extended by Genest and M'Lan [19], but until now, the only publication applied to BSC is the work of Golpîra and Veysi [20], who describe the application to the BSC within a non-profit organization. In this work, the logistic regression and Bradley-Terry method were employed for classifying, sorting, and ranking the factors and finding the most important indexes for establishing the organizational strategy map. Not in the same context as that of BSC but very similar, the Bradley-Terry model has been recently applied for the assessment of environmental driving factors [21]. The authors conducted a study for ranking the parameters for coal-mining activities. In this case, 23 parameters of a coal-mining environment were identified and classified into four major categories, calculating the weight of each parameter using Bradley-Terry. The parameters were ranked by assigning the weights, using attitudinal data collected by surveying experts. More recently, the model has also been applied for measuring portfolios salience [22]. For each matchup between two cabinet portfolios, the subjects (experts or politicians) were asked to choose the more valuable one. They strengthened the greatly simplified data collection of the method. The authors also remark that the application of Bradley-Terry to this context remains rare. There is also another publication where the Bradley-Terry model is used for prioritizing, but it is applied to the design goals of a medical simulator [23], which is far from the operational research field. A survey of pairwise comparisons was distributed to experts. The analysis was performed following two methods: a simple method (calculating the proportion of times an alternative was chosen as preferable) and the Bradley-Terry model. They state that the Bradley-Terry method offers a means to calculate measures of uncertainty, showing nuances where scores may overlap, and the method is valuable when reconciling different experts' opinions. There are no more reported references on the application of the Bradley-Terry model for prioritization in the management domain, even considering that prioritizing performance measures within the balance scorecard is a topic that is very studied nowadays (examples of recent reviews can be consulted in [24–26]). Given the few existing references in the literature, this work contributes by reaffirming the applicability of the method for this purpose and establishing a generalizable framework to be used for BSC prioritization.

3. Materials and Methods

The case study is based on a Spanish software factory that develops software and provides services to several financial entities. The company plays an important role in the information technology sector for financial entities in Spain and South America. The company is a subsidiary firm of a banking group. They have started an important process of adaptation and change of their business model a few years ago, with the goal of improving the efficiency and productivity as a way of ensuring its business sustainability. They have adopted a strategic management approach based on a redefined BSC framework, as published in a former work [27]. Tables 1–4 summarize the BSC KPIs, which includes the four usual perspectives: Financial, Customer, Internal Business Processes, and Learning and Growth. The KPIs were derived from the strategic goals of the organization.

Table 1. Summary of the financial perspective key performance indicators (KPIs).

Code	Name	Description
F1	Cost Structure	Assess the cost evolution in relation to the matrix financial entity size. When the size of the matrix financial group decreases, the costs of the software factory should also decrease in a similar proportion.
F2	Reduction of Cost	The goal of this KPI is to evaluate the percentage of the structural cost of the software factory that is covered by incomes derived by sales to companies outside the corporate group. As a result of the huge cost of software development, sales revenue outside the financial group owner is generally seen as the major reduction of costs.
F3	Useful Developments	Measure the use of the delivered software by the customers. In this particular case, where the company uses a pay-per-use model, the degree of use of the developments is indicated by the number of software executions, and the indicator is calculated as the cumulative number of these executions in relation to the size of the financial institution over the last year. The greater the use, the higher the incomes that should be achieved. It indicates also that the delivered software is useful.

Table 2. Summary of the customer perspective key performance indicators.

Code	Name	Description
C1	User Satisfaction	The indicator measures the degree of customer satisfaction concerning software delivered and services provided by the company. User satisfaction KPI is defined as "the overall level of compliance with the user expectations, measured as a percentage of really met expectations". Therefore, the indicator is an aggregate measure of user satisfaction with various aspects of the service.
C2	Cost per Use	The ratio between the cost paid by the company customers and the degree of use of the provided software. As it is a pay-per-use model, it is measured by means of the cumulative number of executions as in (F3).
C3	Service Level Agreements (SLA)	In the financial software sector, the companies provide critical application services for customers, which need effective mechanisms to manage and control them. SLAs are agreements signed between a service provider and another party such as a service consumer, broker agent, or monitoring agent. The proposed index is a multi-indicator that joins and unifies all the agreements reached with the financial group, and more specifically between the financial institution and the FSF.

Table 3. Summary of the internal business processes' perspective key performance indicators.

Code	Name	Description
I1	Work Performance	This efficiency indicator is calculated as the ratio between budgeted hours and the performed hours.
I2	Employee Productivity	This ratio reflects the amount of software that an employee produces for each hour on the job.
I3	Delay	This indicator shows the average delay in hours.
I4	Software Quality	An aggregated indicator that assesses the company software quality.
I5	Budgeting Error	The indicator shows how good the estimations were over the last year.

Table 4. Summary of the learning and growth perspective key performance indicators.

Code	Name	Description
L1	Employer Branding	Reputation of the firm as an employer. The most important metrics are employee satisfaction, employee engagement and loyalty, quality of hire, time and cost per hire, job acceptance rate of candidates, number of applicants, employee turnover, increased level of employee referrals, decreased absenteeism, promotion readiness rating, external/internal hire ratio, performance ratings of newly promoted managers, and manager/executive failure rate.
L2	Intellectual Capital	An aggregated indicator that assesses the intellectual capital as a compendium of human, structural, and relational capital.

This framework has been preferred among other existing frameworks in the literature [28] because it has been designed tailored to the environment of this kind of FSF and is the one established in the studied company. Some of the KPIs are simple (i.e., F2-Reduction of cost), but others are complex (i.e., C1-User Satisfaction, C3-SLA, I4-Software Quality, or L1-Employer Branding) because they group several sub-indicators. The description of the KPIs included in the BSC, the method used to measure every KPI, and their justification are extensively explained in [27].

3.1. Analytic Hierarchy Process Method

The method was devised by Saaty in the 1970s [3] and it has been adopted as one of the most used MCDM processes until now. The method is used to prioritize the relative importance of criteria by making pairwise comparisons, instead of sorting, voting, or freely assigning priorities. Saaty establishes an intensity of importance for the comparisons on an absolute scale with nine levels, which is known as Saaty's scale [29]. The method starts with defining the goal of the decision and the alternatives and structuring them in a hierarchy. Then, the pairwise comparison of criteria in each category is performed, and the priorities are derived.

If an alternative A_i is preferable to an alternative A_j, then the value of the comparison scale $P_c(A_i, A_j) = a_{ij}$ indicates the intensity of relative importance of A_i over A_j. The matrix A is the result of all of the comparisons and represents the relative importance a_{ij} of each element.

The method uses the principal eigenvalue method to derive the priorities. The calculation of weights relies on an iterative process in which matrix A is successively multiplied by itself, resulting in normalized weights, w_i, which represents the importance of alternative A_i relative to all other alternatives.

The judgment of decision-makers in pairwise comparisons may present inconsistencies when all of the alternatives are taken into consideration simultaneously. So, the consistency index (CI) and the consistency ratio (CR) are calculated to measure the degree to which judgments are not coherent [30]. It is normally considered that if CR < 0.10, then the degree of consistency is satisfactory [31]. If the maximum eigenvalue, CI, and CR are satisfactory, then a decision is taken based on the normalized values; else, the procedure is repeated until these values lie in a desired range.

A good description of the usage and different applications of AHP can be found at the work of Ishizaka and Labib [32]. The evolution of the method can be also followed in the Emrouznejad and Marra publication [33]. The application related to this work is the utilization of AHP for priority and ranking, where it has been extensively used.

The application of AHP it is not always easy. The number of comparisons grows exponentially according to the number different criterion to be considered. The scale presents some difficulties also, being subjective for the decision-makers discerning between the different levels of intensity of importance when comparing two alternatives. As Buckley and Uppuluri [34] remark, "It is difficult for people to always assign exact ratios when comparing two alternatives." In a similar way, Chang [35] states that, "Due to the complexity and uncertainty involved in real world decision problems, it is sometimes unrealistic or even impossible to require exact judgments." In addition, the consistency analysis is complicated when a large number of decision-makers are involved, resulting in a complex post-processing process that could entail leaving out several opinions. In addition, despite its wide use, the method is not free of criticism from various perspectives. For example, Costa and Vansnick [36] state that the priority vector derived can violate the so-called "condition of order preservation" that is fundamental in decision-making.

For a long period, the predominant tendency was to extend the method by hybridizing it with other methods and thus introducing a higher complexity. The original method was combined with Fuzzy Set theory [37] given the Fuzzy Analytic Hierarchy Process (FAHP) method [35]. Regardless, the introduced complexity of these new methods (more complex questionnaires, fuzzification and defuzzification models, complexity when calculation, and difficulty of interpretation of the results), there is some controversy about the real benefits. For example, the paper published by K. Zhü openly criticizes the fuzzy approaches to AHP [38]. The author claims that despite the popularity of the method, this approach has problems, stating that the operational rules of fuzzy numbers oppose the logic of the AHP and analyzing the validity, among other things. K. Zhü holds the opinion that, "It is not necessary to use a complex paradigm to express complex things, sometimes a simple paradigm may be better." Thomas L. Saaty has also paid close attention to these extensions, writing some papers from a critical perspective [39,40]. By contrast, a tendency has recently emerged trying to simplify the application of the method as much as possible. For example, Leal [41] develops a simplified method that calculates the priorities of each alternative against a set of criteria with only $n - 1$ comparisons of n alternatives for each criterion, instead of $n*(n - 1)/2$ comparisons in the original method.

In any case, our study does not concern the validity or not of the AHP and its extensions, but we want to emphasize the idea that the simplest methods under certain conditions are the most appropriate. Under this context, the Bradley-Terry model could be an easier method, as the Saaty's scale could be transformed in a win-to-loss scale (the decision-makers only need to specify which criterion is preferred in the comparison, without grading the intensity of importance), and the computing of data might be simpler.

3.2. Bradley-Terry Model and BSC Integration

The Bradley-Terry model [42] is a method of analysis of paired comparisons based on the logit model. A general introduction can be found in Agresti [9]. Given a pair of individuals i and $j / (i,j \in \{1, \ldots, K\})$, the model estimates the probability that i is preferred to j as:

$$P(i > j) = \frac{p_i}{p_i + p_j}. \tag{1}$$

In the Expression (1), p_i is a positive real-valued score (the underlying worth of each item) assigned to individual i and $P(i > j) + P(j > i) = 1$ for all of the pairs. The Bradley-Terry model uses exponential score functions, so the probability of selection is expressed in terms of exponential functions:

$$p_i = e^{\beta_i} \tag{2}$$

Thus, Expression (1) can be expressed as:

$$P(i > j) = \frac{e^{\beta_i}}{e^{\beta_i} + e^{\beta_j}}. \tag{3}$$

Alternatively, it can be expressed using the logit as:

$$\text{logit}(P(i > j)) = \log\left(\frac{P(i > j)}{1 - P(i > j)}\right) = \log\left(\frac{P(i > j)}{P(j > i)}\right) = \beta_i - \beta_j. \tag{4}$$

Then, the parameters $\{p_i\}$ can be estimated by maximum likelihood using the Zermelo [43] method. Standard software for generalized linear models can be used for the computing as described by Turner and Firth [44], who are the authors of one of the most used packages for Bradley-Terry calculation under R software.

The observations required are the outcomes of previous comparisons, which are expressed as pairs (i,j), counting the number of times that i is preferred to j and summarizing these outcomes as w_{ij}. Thus, w_{ij} accounts the times that an indicator i was preferred to j by the decision-makers. The log-likelihood of $\{p_i\}$ can be obtained as:

$$\ell(p) = \sum_{i=1}^{m} \sum_{j=1}^{m} \left[w_{ij} \ln(p_i) - w_{ij} \ln(p_i + p_j) \right]. \tag{5}$$

It is assumed by convention that $w_{ii} = 0$. Starting from an arbitrary vector p, the algorithm iteratively performs the update

$$p'_i = W_i \left(\sum_{i \neq j} \frac{w_{ij} + w_{ji}}{p_i + p_j} \right)^{-1} \tag{6}$$

for all i, where W_i is the number of comparisons 'won' by i. After computing all the parameters, they should be renormalized, so $\sum_i p_i = 1$.

Additional extensions have been proposed. For example Böckenholt [45] proposed a method for ranking more than two options. The model has also been extended to allow ordinal comparisons. In this case, the subjects can make their preference decisions on more than two preference categories. The works of Tutz [46], Agresti [47], Dittrich et al. [48], and Casalicchio et al. [49] provide extensions in this sense. However, they are unnecessary here, because our goal is to retrieve the underlying relative worth of each indicator in a simple way. For the calculation, statistical packages have been developed and described in the literature, most of them R extensions: Firth [50], Turner and Firth [44], Hankin [51], or Clark [52], for example.

In terms of calculation, the process starts surveying the decision-makers through a pairwise questionnaire. The difference with respect to the AHP method is that Saaty's scale is not used. Instead, they indicate which indicator is the most important (the 'winner'), without expressing a degree of preference. Then, a table is built summarizing the number of times each indicator 'wins'. For example, in the case of 4 KPIs, the table will follow the structure shown in Table 5:

Table 5. Data aggregation example for the Bradley-Terry calculation in a win-to-loss context.

Factor 1	Factor 2	Win1	Win2
KPI_1	KPI_2	N_{12}	N_{21}
KPI_1	KPI_3	N_{13}	N_{31}
KPI_1	KPI_4	N_{14}	N_{41}
KPI_2	KPI_3	N_{23}	N_{32}
KPI_2	KPI_4	N_{24}	N_{42}
KPI_3	KPI_4	N_{34}	N_{43}

Here, N_{ij} stands for the number of times KPI_i was preferred to KPI_j. This is a form of coding widely used by most R extensions that allows the calculations of the Bradley-Terry model.

3.3. Method and Empirical Application

A demonstration of the method is explained in this section. It is necessary to bear in mind that in this case, we have started from the data collected in our former study [53]. The entire process is enumerated, although only those steps specific of the Bradley-Terry modeling are presented in detail. The steps taken to achieve this purpose are:

1. Analyze the BSC of the studied organization.

2. Define the hierarchical framework according to each perspective of the BSC.
3. Survey the decision-makers' opinions regarding the indicators and perspectives of the BSC using a pairwise questionnaire in a win-to-loss context.
4. Prepare the answers to be processed with Bradley-Terry software.
5. Compute the perspectives and indicators' weights.
6. Rank the indicators.
7. Analyze the results and obtain conclusions.

The process is depicted in Figure 1.

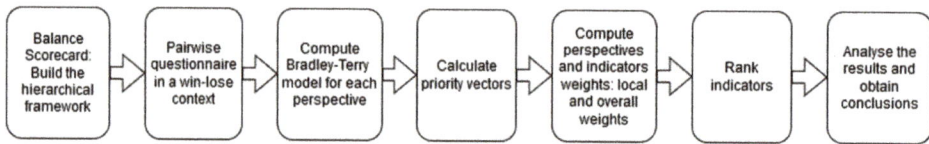

Figure 1. Process framework.

Figure 2 shows the hierarchical model of the BSC. The specific set of 13 KPIs (Tables 1–4) are grouped according to their related perspective.

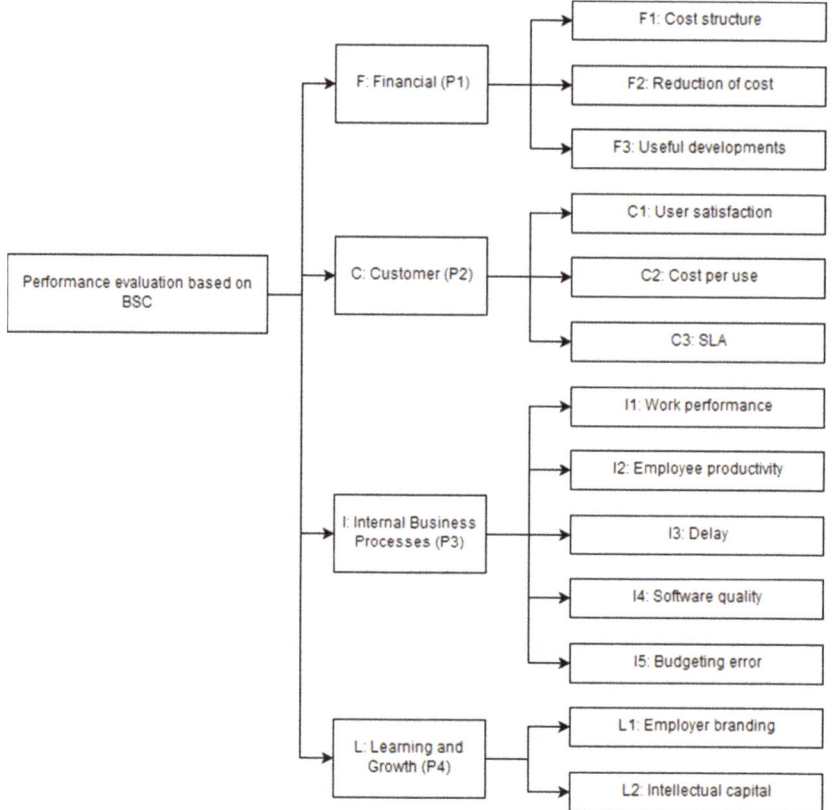

Figure 2. Hierarchical framework of Balance Scorecard (BSC) performance evaluation criteria for a Financial Software Factory (FSF). Adapted from [27,53].

According to the hierarchical structure shown in Figure 2, a conventional questionnaire in AHP format was distributed. The questionnaires were sent to different internal and external stakeholders of the company and some experts in the field of software factories to ask for their professional point of view on sustainability and performance goals in relation to the company scenario. The number of questionnaires sent was 83, and the number of received questionnaires was 61, which represents 73% of the total of questionnaires sent. A detailed description of the considered roles and additional details about the survey process can be found in the section "Data Collection" from our former paper, where the AHP prioritization was published [53].

In this particular case, we have started from an AHP conventional questionnaire, following the scale proposed by Saaty [29], but we have transformed the answers into a win-to-loss context for the application of our method. For each comparison, we have considered only who is the winner (the preferred factor by the expert in the comparison) and the loser, regardless of the intensity of the preference. We have considered the special case of equal importance as a tie. It must be taken into consideration that our proposal starting from scratch consists of carrying out a questionnaire without considering intensities, simply forcing the respondent to indicate the most important factor in the pairwise comparison (or equal).

As a consequence, five different files were built, as shown in Tables 6–10. Table 6 shows decision-makers' preferences regarding the four different perspectives, which are denoted as P1, ... , P4. The column Win1 denotes the number of times that Factor 1 was preferred over Factor 2. For example, the first row in Table 6 indicates that P1 (Financial Perspective) was preferred by 13 of the respondents over P2 (Customer Perspective), and P2 was preferred by 48 of the respondents over P1. The particular case when the respondent has indicated equal importance is considered as a tie, and then a half point is assigned to each factor, truncating the result to an integer number in order to be computed by the Bradley-Terry model.

Table 6. Input file of Perspectives (P) preferences in a win-to-loss context.

Factor 1	Factor 2	Win 1	Win 2
P1	P2	13	48
P1	P3	34	27
P1	P4	31	29
P2	P3	51	10
P2	P4	50	10
P3	P4	32	29

Table 7. Input file of Financial (F) factors preferences in a win-to-loss context.

Factor 1	Factor 2	Win 1	Win 2
F1	F2	41	19
F1	F3	29	31
F2	F3	26	34

Table 8. Input file of Customer (C) factors preferences in a win-to-loss context.

Factor 1	Factor 2	Win 1	Win 2
C1	C2	53	7
C1	C3	39	22
C2	C3	20	41

Table 9. Input file of Internal (I) factors preferences in a win-to-loss context.

Factor 1	Factor 2	Win 1	Win 2
I1	I2	43	18
I1	I3	36	24
I1	I4	21	40
I1	I5	38	23
I2	I3	33	29
I2	I4	19	42
I2	I5	37	23
I3	I4	15	45
I3	I5	39	22
I4	I5	51	10

Table 10. Input file of Learning and Growth (L) factors preferences in a win-to-loss context.

Factor 1	Factor 2	Win1	Win2
L1	L2	18	42

The data was processed using the extension "BradleyTerry2" for R, following the process described by Turner and Firth [44]. RStudio version 1.2.1335 was used for the computation with a standard Core i5 computer. The standard Bradley-Terry model was used alongside fitting by maximum likelihood. The coefficients returned by the model ($\hat{\beta}_i$) are the model estimations setting $\hat{\beta}_0 = 0$. In order to turn these coefficients into the BSC weights w_i, they must be transformed calculating $exp(\hat{\beta}_i)$ and normalizing the setting $\sum_i (\hat{\beta}_i) = 1$. The results for the BSC perspectives are presented in Table 11 as an example.

Table 11. Results of fitting the Bradley-Terry model to Perspectives data.

Factor	$\hat{\beta}_i$	$exp(\hat{\beta}_i)$	w_i
P1	0	1	0.1493
P2	1.3917	4.0217	0.6006
P3	−0.1740	0.8403	0.1255
P4	−0.1809	0.8345	0.1246

We have all the local weights of the indicators after computing the Bradley-Terry model for each set (Tables 7–10), denoted w_{Pij} (the weight of the indicator j belonging to the perspective i). The next step is to calculate the overall weights of the sub-criteria, W_{Pij}. The local weight of each sub-criteria is multiplied by its corresponding relative importance of the criteria (w_{Pi}). Mathematically, it can be expressed as given in Equation (7). The overall weight is finally used for ranking the indicators.

$$W_{Pij} = w_{Pi} * w_{Pij} \qquad (7)$$

4. Results and Discussion

The results after processing all the data are presented in Tables 12 and 13. The number of considered questionnaires that passed the consistency test when computing the AHP was 44. In order to compare results, the Bradley-Terry model has been calculated in two different ways: computing the 61 questionnaires (denoted as Bradley-Terry-61) and computing the 44 questionnaires (denoted as Bradley-Terry-44) that have passed the AHP consistency test.

Table 12 shows the local weights for the computed Bradley-Terry model compared with the local results provided by the application of the AHP [53]. Table 13 presents the overall weights as well as the ranking of each one.

Table 12. Bradley-Terry local weights compared with the Analytic Hierarchy Process method (AHP).

Criteria and Sub-Criteria	Bradley-Terry-61	Bradley-Terry-44	AHP
(F) Financial	0.1493	0.1779	0.2035
(F1) Cost Structure	0.4066	0.4415	0.4134
(F2) Reduction of Cost	0.2304	0.2230	0.2438
(F3) Useful Developments	0.3630	0.3355	0.3429
(C) Customer	0.6006	0.5704	0.4586
(C1) User Satisfaction	0.6032	0.6455	0.5411
(C2) Cost per Use	0.1137	0.1047	0.1712
(C3) SLA	0.2831	0.2498	0.2876
(I) Internal Business Processes	0.1255	0.1291	0.1712
(I1) Work Performance	0.2207	0.2365	0.2118
(I2) Employee Productivity	0.1423	0.1345	0.1516
(I3) Delay	0.1424	0.1658	0.1587
(I4) Software Quality	0.4005	0.3753	0.3698
(I5) Budgeting Error	0.0941	0.0879	0.1082
(L) Learning and Growth	0.1246	0.1226	0.1667
(L1) Employer Branding	0.3000	0.3488	0.3708
(L2) Intellectual Capital	0.7000	0.6512	0.6292

Table 13. Bradley-Terry overall weights and rank compared with AHP.

Criteria and Sub-Criteria	Bradley-Terry-61		Bradley-Terry-44		AHP	
	Weights	Rank	Weights	Rank	Weights	Rank
(F) Financial						
(F1) Cost Structure	0.0607	5	0.0785	4	0.0841	4
(F2) Reduction of Cost	0.0344	9	0.0397	9	0.0496	9
(F3) Useful Developments	0.0542	6	0.0597	6	0.0698	6
(C) Customer						
(C1) User Satisfaction	0.3623	1	0.3682	1	0.2482	1
(C2) Cost per Use	0.0683	4	0.0597	5	0.0785	5
(C3) SLA	0.1700	2	0.1425	2	0.1319	2
(I) Internal Business Processes						
(I1) Work Performance	0.0277	10	0.0305	10	0.0363	10
(I2) Employee Productivity	0.0179	12	0.0174	12	0.0260	12
(I3) Delay	0.0179	11	0.0214	11	0.0272	11
(I4) Software Quality	0.0503	7	0.0484	7	0.0633	7
(I5) Budgeting Error	0.0118	13	0.0114	13	0.0185	13
(L) Learning and Growth						
(L1) Employer Branding	0.0374	8	0.0428	8	0.0618	8
(L2) Intellectual Capital	0.0872	3	0.0798	3	0.1049	3

The results indicate that the "Customer Perspective" is the main point of attention, followed by the "Finance Perspective". The "Learning and Growth" and "Internal Processes" perspectives, almost with the same weights, are the least considered. The rank remains unchanged regardless of the method used. As can be noted from Table 12, the weights are quite similar for every indicator.

Regarding the indicators, the differences are not significant. The indicators "User Satisfaction" and "SLA" are the most rated for all the methods, followed by the "Intellectual Capital", Taking the AHP weights as reference, the mean square error considering Bradley-Terry with 61 questionnaires is 0.01 and considering 44 questionnaires is 0.001. In addition, the only difference in the rank is one position between F1 (Cost Structure) and C2 (Cost per use), as remarked with bold font in Table 13.

Figure 3 helps to visualize the different weights of the models for each indicator. As it can be noticed, there are few differences. The most notable difference is that the Bradley-Terry models increase the weight of the C1 indicator (User Satisfaction) compared to AHP.

Therefore, in view of the results, it is shown that the proposed method can be used as an alternative to AHP. One of the main advantages is the simplification of the scale ('win-to-loss' context instead of the traditional 9-point scale), which can be an advantage when the decision-makers have to make many comparisons and run the risk of losing the rigor. Another advantage derived from the previous is that the level of inconsistency is more reduced. In addition, the calculation of the Bradley-Terry model is tolerant to missing comparisons; therefore, to some extent, the comparisons could be reduced or it could be accepted that decision-makers do not answer all the comparisons.

The main limitation derives from the fact that the method can only be applied in a context of group consensus among several decision-makers. This should not be a problem in the applied field, since the prioritization of indicators is always carried out with the idea of combining different opinions. However, it remains unclear whether there is a minimum number of participants required for the application of the method. A further study could be performed to investigate more about this aspect. In addition, another limitation is that the method does not incorporate any mechanism to check for the consistency. In the case studied, the method provided reliable results even for the data that had not passed the AHP consistency test (BTM-61), but it is unclear whether the behavior works against more severe levels of inconsistency in the responses. The search for an inconsistency index that is applicable to win-to-loss pairwise comparisons is proposed as further work. The survey done by Brunelli [54] could constitute a good starting point. The author appoints several methods for different representations of pairwise comparisons and also details how to deal with group decision-making. Once a valid index is identified, a comparison of the results from the different methods could be performed in a similar way to the analysis done by Genest and M'Lan [19]. Finally, another limitation of this study is that it has been validated only in the exposed case. The case is very general and representative, but as mentioned, aspects not studied such as the sensitivity to the degree of inconsistency or the number of responses might generate uncertainty.

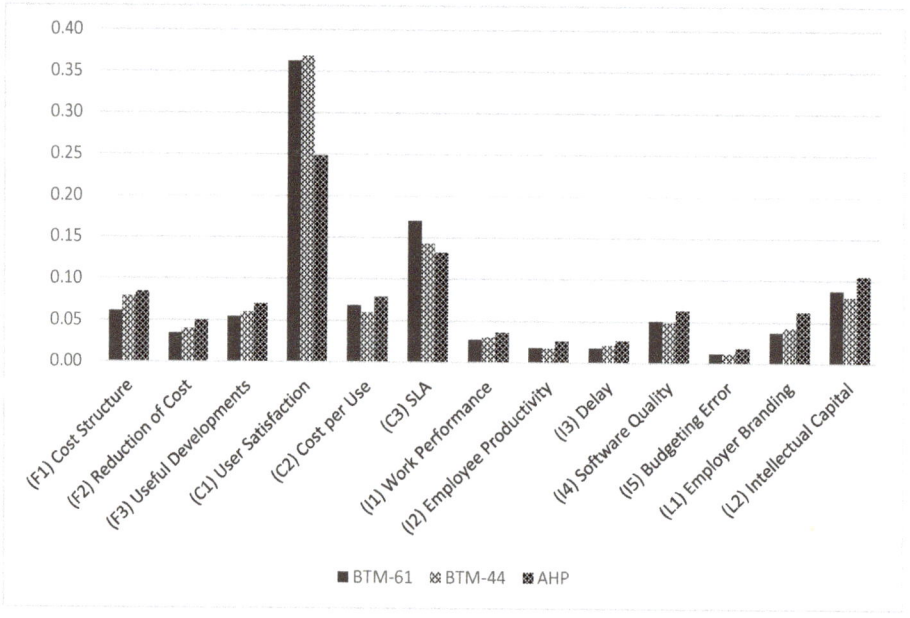

Figure 3. Sub-criteria (KPIs) weights chart comparing the models.

5. Conclusions

The case of how to determine the weights of the BSC KPIs based on the Bradley-Terry model is presented here. The method, compared with a traditional application of AHP, provides quite similar results while simplifying the whole process, even more if we consider more complicated variations of AHP such as for example Fuzzy AHP. This aligns with the statements of other authors (i.e., Zucco Jr et al. [22] and Clark et al. [52]).

The scale for the comparisons was simplified regarding the usual AHP scale, considering only the winner and the loser of the pairwise comparisons, accepting ties also as an option. This is an important advantage when decision-makers must make the assessments, since it simplifies comparisons, especially when there are many factors to consider.

The method exposed also has the advantage of simplifying the calculation process by not having to evaluate the consistency ratio. The consistency test procedure followed with AHP involves usually analyzing those decision-makers' answers that cause the consistency ratio to fall below the limits. In the AHP base case, the answers from 17 decision-makers had not been considered because of this effect. This could be a significant issue in surveys with few decision-makers.

Based on the experience implementing this model within the studied company, we could remark as an important conclusion that what really matters is not the exact weight of each indicator but rather the general ranking of indicators. Under this situation, we can state that using a more simplified method such as the one presented here does not provide significant differences regarding the ranking of indicators. So, when planning the deployment of a performance management system, it should be considered whether using a more complex method such as AHP is worthwhile.

This paper contributes to the multi-criteria decision-making domain reporting a successful application of the Bradley-Terry model for weighting the BSC with a simplified scale for the pairwise comparisons and confronting against the AHP results, which is something that has been barely documented in the literature. The method was applied to the case of a Spanish software company, but the approach could be extrapolated to any organization that presents a similar framework to the one exposed. The combination of AHP methods with BSC has been demonstrated in the literature to be a very valuable tool for performance evaluation and making strategic decisions. However, comparing the results obtained using AHP with the Bradley-Terry model, we believe that the AHP does not add extra value in this situation; meanwhile, the calculation is slightly more complex.

Author Contributions: Conceptualization, V.R.M. and C.Á.P.; formal analysis, J.M.V.B. and M.D.P.; investigation, V.R.M. and C.Á.P.; methodology, V.R.M.; supervision, V.R.M.; writing—original draft, V.R.M.; writing—review and editing, M.D.P. and J.M.V.B. All authors have read and agreed to the published version of the manuscript.

Funding: This work has been subsidized through the Plan of Science, Technology and Innovation of the Principality of Asturias (Ref: FC-GRUPIN-IDI/2018/000225).

Conflicts of Interest: The authors declare no conflict of interest.

References

1. Kaplan, R.S.; Norton, D.P. The balanced scorecard—Measures that drive performance. *Harv. Bus. Rev.* **1992**, *70*, 71–79. [PubMed]
2. Kaplan, R.S.; Norton, D.P. Putting the balanced scorecard to work. *Harv. Bus. Rev.* **1993**, *71*, 134–140.
3. Saaty, T.L. A scaling method for priorities in hierarchical structures. *J. Math. Psychol.* **1977**, *15*, 234–281. [CrossRef]
4. Clinton, B.D.; Webber, S.A.; Hassell, J.M. Implementing the balanced scorecard using the analytic hierarchy process. *Manag. Account. Q.* **2002**, *3*, 1–11.
5. Reisinger, H.; Cravens, K.S.; Tell, N. Prioritizing performance measures within the balanced scorecard framework. *Manag. Int. Rev.* **2003**, *43*, 429.
6. Vaidya, O.S.; Kumar, S. Analytic hierarchy process: An overview of applications. *Eur. J. Oper. Res.* **2006**, *169*, 1–29. [CrossRef]

7. Saaty, T.L. On the measurement of intengibles. A principal eigenvector approach to relative measurement derived from paired comparisons. *Not. Am. Math. Soc.* **2013**, *60*, 192–208. [CrossRef]
8. Leung, L.C.; Cao, D. On consistency and ranking of alternatives in fuzzy AHP. *Eur. J. Oper. Res.* **2000**, *124*, 102–113. [CrossRef]
9. Agresti, A. *Categorical Data Analysis*; John Wiley & Sons: Hoboken, NJ, USA, 2003; Volume 482, ISBN 0-471-45876-7.
10. McHale, I.; Morton, A. A Bradley-Terry type model for forecasting tennis match results. *Int. J. Forecast.* **2011**, *27*, 619–630. [CrossRef]
11. Koehler, K.J.; Ridpath, H. An application of a biased version of the Bradley-Terry-Luce model to professional basketball results. *J. Math. Psychol.* **1982**, *25*, 187–205. [CrossRef]
12. Király, F.J.; Qian, Z. Modelling Competitive Sports: Bradley-Terry- Élő Models for Supervised and On-Line Learning of Paired Competition Outcomes. *arXiv* **2017**, arXiv:170108055.
13. Stigler, S.M. Citation patterns in the journals of statistics and probability. *Stat. Sci.* **1994**, 94–108. [CrossRef]
14. Courcoux, P.; Semenou, M. Preference data analysis using a paired comparison model. *Food Qual. Prefer.* **1997**, *8*, 353–358. [CrossRef]
15. Loewen, P.J.; Rubenson, D.; Spirling, A. Testing the power of arguments in referendums: A Bradley-Terry approach. *Elect. Stud.* **2012**, *31*, 212–221. [CrossRef]
16. Fienberg, S.E.; Meyer, M.M. Loglinear models and categorical data analysis with psychometric and econometric applications. *J. Econom.* **1983**, *22*, 191–214. [CrossRef]
17. Cattelan, M. Models for paired comparison data: A review with emphasis on dependent data. *Stat. Sci.* **2012**, 412–433. [CrossRef]
18. Dykstra, O. Rank analysis of incomplete block designs: A method of paired comparisons employing unequal repetitions on pairs. *Biometrics* **1960**, *16*, 176–188. [CrossRef]
19. Genest, C.; M'lan, C.-É. Deriving priorities from the Bradley-Terry model. *Math. Comput. Model.* **1999**, *29*, 87–102. [CrossRef]
20. Golpîra, H.; Veysi, B. Flexible balanced Scorecard for nonprofit organizations. *Adv. Ind. Eng. Inf. Water Resour.* **2012**, 139–146.
21. Bhar, C.; Srivastava, V. Environmental capability: A Bradley-Terry model-based approach to examine the driving factors for sustainable coal-mining environment. *Clean Technol. Environ. Policy* **2018**, *20*, 995–1016.
22. Zucco, C., Jr.; Batista, M.; Power, T.J. Measuring portfolio salience using the Bradley-Terry model: An illustration with data from Brazil. *Res. Polit.* **2019**, *6*, 2053168019832089.
23. Dorton, S.; Frommer, I.; Bailey, M.; Sotomayor, T. Prioritizing Design Goals for a Medical Simulator Using Pairwise Comparisons. In Proceedings of the Human Factors and Ergonomics Society Annual Meeting, Philadelphia, PA, USA, 1–5 October 2018; SAGE Publications Sage CA: Los Angeles, CA, USA, 2018; Volume 62, pp. 1648–1652.
24. Janíčková, N.; Žižlavský, O. Key performance indicators and the Balanced Scorecard approach in small and medium-sized enterprises: A literature review. In Proceedings of the International Conference at Brno University of Technology—Faculty of Business and Management, Brno, Czech Republic, 30 April 2019.
25. Quesado, P.R.; Aibar Guzmán, B.; Lima Rodrigues, L. Advantages and contributions in the balanced scorecard implementation. *Intang. Cap.* **2018**, *14*, 186–201. [CrossRef]
26. Janeš, A.; Kadoić, N.; Begičević Ređep, N. Differences in prioritization of the BSC's strategic goals using AHP and ANP methods. *J. Inf. Organ. Sci.* **2018**, *42*, 193–217. [CrossRef]
27. Álvarez, C.; Rodríguez, V.; Ortega, F.; Villanueva, J. A Scorecard Framework Proposal for Improving Software Factories' Sustainability: A Case Study of a Spanish Firm in the Financial Sector. *Sustainability* **2015**, *7*, 15999–16021. [CrossRef]
28. Peredo Valderrama, R.; Canales Cruz, A.; Peredo Valderrama, I. An Approach Toward a Software Factory for the Development of Educational Materials under the Paradigm of WBE. *Interdiscip. J. E-Learn. Learn. Objects* **2011**, *7*, 55–67. [CrossRef]
29. Saaty, T.L. How to make a decision: The analytic hierarchy process. *Eur. J. Oper. Res.* **1990**, *48*, 9–26. [CrossRef]
30. Sharma, M.K.; Bhagwat, R. An integrated BSC-AHP approach for supply chain management evaluation. *Meas. Bus. Excell.* **2007**, *11*, 57–68. [CrossRef]

31. Saaty, T.L. An exposition of the AHP in reply to the paper "remarks on the analytic hierarchy process". *Manag. Sci.* **1990**, *36*, 259–268. [CrossRef]
32. Ishizaka, A.; Labib, A. Review of the main developments in the analytic hierarchy process. *Expert Syst. Appl.* **2011**, *38*, 14336–14345. [CrossRef]
33. Emrouznejad, A.; Marra, M. The state of the art development of AHP (1979–2017): A literature review with a social network analysis. *Int. J. Prod. Res.* **2017**, *55*, 6653–6675. [CrossRef]
34. Buckley, J.J.; Uppuluri, V.R.R. Fuzzy hierarchical analysis. In *Uncertainty in Risk Assessment, Risk Management, and Decision Making*; Springer: New York, NY, USA, 1987; pp. 389–401.
35. Chang, D.-Y. Applications of the extent analysis method on fuzzy AHP. *Eur. J. Oper. Res.* **1996**, *95*, 649–655. [CrossRef]
36. Costa, C.A.B.; Vansnick, J.-C. A critical analysis of the eigenvalue method used to derive priorities in AHP. *Eur. J. Oper. Res.* **2008**, *187*, 1422–1428. [CrossRef]
37. Zadeh, L.A. Fuzzy sets. *Inf. Control* **1965**, *8*, 338–353. [CrossRef]
38. Zhü, K. Fuzzy analytic hierarchy process: Fallacy of the popular methods. *Eur. J. Oper. Res.* **2014**, *236*, 209–217. [CrossRef]
39. Saaty, T.L. There is no mathematical validity for using fuzzy number crunching in the analytic hierarchy process. *J. Syst. Sci. Syst. Eng.* **2006**, *15*, 457–464. [CrossRef]
40. Saaty, T.L.; Tran, L.T. On the invalidity of fuzzifying numerical judgments in the Analytic Hierarchy Process. *Math. Comput. Model.* **2007**, *46*, 962–975. [CrossRef]
41. Leal, J.E. AHP-express: A simplified version of the analytical hierarchy process method. *MethodsX* **2019**, *7*, 100748. [CrossRef]
42. Bradley, R.A.; Terry, M.E. Rank analysis of incomplete block designs: I. The method of paired comparisons. *Biometrika* **1952**, *39*, 324–345. [CrossRef]
43. Zermelo, E. Die berechnung der turnier-ergebnisse als ein maximumproblem der wahrscheinlichkeitsrechnung. *Math. Z.* **1929**, *29*, 436–460. (In German) [CrossRef]
44. Turner, H.; Firth, D. Bradley-Terry models in R: The BradleyTerry2 package. *J. Stat. Softw.* **2012**, *48*, 1–21. [CrossRef]
45. Böckenholt, U. Hierarchical modeling of paired comparison data. *Psychol. Methods* **2001**, *6*, 49. [CrossRef] [PubMed]
46. Tutz, G. Bradley-Terry-Luce models with an ordered response. *J. Math. Psychol.* **1986**, *30*, 306–316. [CrossRef]
47. Agresti, A. Analysis of ordinal paired comparison data. *J. R. Stat. Soc. Ser. C Appl. Stat.* **1992**, *41*, 287–297. [CrossRef]
48. Dittrich, R.; Francis, B.; Hatzinger, R.; Katzenbeisser, W. A paired comparison approach for the analysis of sets of Likert-scale responses. *Stat. Model.* **2007**, *7*, 3–28. [CrossRef]
49. Casalicchio, G.; Tutz, G.; Schauberger, G. Subject-specific Bradley-Terry-Luce models with implicit variable selection. *Stat. Model.* **2015**, *15*, 526–547. [CrossRef]
50. Firth, D. Bradley-Terry models in R. *J. Stat. Softw.* **2005**, *12*, 1–12. [CrossRef]
51. Hankin, R.K. Partial Rank Data with the hyper2 Package: Likelihood Functions for Generalized Bradley-Terry Models. *R J.* **2017**, *9*, 429–439. [CrossRef]
52. Clark, A.P.; Howard, K.L.; Woods, A.T.; Penton-Voak, I.S.; Neumann, C. Why rate when you could compare? Using the "EloChoice" package to assess pairwise comparisons of perceived physical strength. *PLoS ONE* **2018**, *13*, e0190393. [CrossRef]
53. Álvarez Pérez, C.; Rodríguez Montequín, V.; Ortega Fernández, F.; Villanueva Balsera, J. Integrating Analytic Hierarchy Process (AHP) and Balanced Scorecard (BSC) Framework for Sustainable Business in a Software Factory in the Financial Sector. *Sustainability* **2017**, *9*, 486. [CrossRef]
54. Brunelli, M. A survey of inconsistency indices for pairwise comparisons. *Int. J. Gen. Syst.* **2018**, *47*, 751–771. [CrossRef]

© 2020 by the authors. Licensee MDPI, Basel, Switzerland. This article is an open access article distributed under the terms and conditions of the Creative Commons Attribution (CC BY) license (http://creativecommons.org/licenses/by/4.0/).

Article
A Novel Coordinated TOPSIS Based on Coefficient of Variation

Pengyu Chen

School of Geography & Resource Science, Neijiang Normal University, Neijiang 641100, China; chenpengyu@njtc.edu.cn; Tel.: +86-0832-2340771

Received: 22 June 2019; Accepted: 9 July 2019; Published: 11 July 2019

Abstract: Coordinated Technique for Order Preference by Similarity to Ideal Solution (TOPSIS) is a significant improvement of TOPSIS, which take into account the coordination level of attributes in the decision-making or assessment. However, in this study, it is found that the existing coordinated TOPSIS has some limitations and problems, which are listed as follows. (1) It is based on modified TOPSIS, not the original TOPSIS. (2) It is inapplicable when using vector normalization. (3) The calculation formulas of the coordination degree are incorrect. (4) The coordination level of attributes is interrelated with the weights. In this paper, the problems of the existing coordinated TOPSIS are explained and revised, and a novel coordinated TOPSIS based on coefficient of variation is proposed to avoid the limitations. Comparisons of the existing, revised, and proposed coordinated TOPSIS are carried out based on two case studies. The comparison results validate the feasibility of the proposed coordinated TOPSIS.

Keywords: TOPSIS; coordinated TOPSIS; decision-making; assessment; coefficient of variation; information entropy

1. Introduction

The Technique for Order Preference by Similarity to Ideal Solution (TOPSIS) is a classical multi-attribute decision-making method, which was first put forward by Hwang and Yoon in 1981 [1]. This method attempts to rank the alternatives by calculating their distances from the ideal solution (IS) and the negative ideal solution (NIS) and selects the optimum one that should simultaneously have the shortest distance from the IS and the farthest distance from the NIS. TOPSIS has been successfully applied in various fields [2–4] such as environmental risk assessment [5], water quality assessment [6], disaster risk management [7,8], supplier selection [9], real estate management [10,11], and sustainability assessment [12] due to the fact that it features a simple principle, easy understanding, and strong capacity to integrate other methods.

Although TOPSIS has many advantages, many scholars have found that it has some limitations for application and put forward some improvement approaches. For example, when alternatives are described by statistically connected criteria, the application of TOPSIS may cause improper ranking results [13]. To avoid this problem, Antuchevičienė et al. (2010) suggested using the Mahalanobis distance instead of Euclidean distance for TOPSIS [13]. Yang and Wu (2019) pointed out that TOPSIS does not consider the data distribution of the degree of dispersion and aggregation when it is compared with the IS and the NIS and proposed a novel TOPSIS based on improved grey relational analysis [14]. Chen's research suggests that when the alternatives are added or reduced, the problem of rank reversal will occur. To void this problem, the absolute IS and the absolute NIS should be used [15]. Considering that TOPSIS lacks evaluation from the perspective of attribute coordination, Yu et al. (2018) proposed coordinated TOPSIS, which takes into account the coordination level of attributes [16].

TOPSIS is a type of multi-attribute decision-making method, in which complementarity exists among attributes. When some attributes are inferior, they can be complemented by the other superior attributes in the decision results [16]. However, in some decision or evaluation problems, the coordination or balance of attributes is very important to be considered, such as evaluation of information system [17], decision of real estate location [18], and evaluation of academic journals [16]. In these cases, the coordination level of attributes should be taken into account when using TOPSIS. Coordinated TOPSIS is a significant improvement of TOPSIS. It is very suitable for the decision or evaluation fields that need to consider the coordination level of attributes [16]. However, according to the principles of coordinated TOPSIS [16], it is found that this method has some limitations and problems that affect its applicability. In this study, the problems of the existing coordinated TOPSIS are explained and revised, and a novel coordinated TOPSIS based on coefficient of variation is proposed to avoid the limitations. The rest of this paper is organized as follows: Section 2 describes the methods; Section 3 gives the results and discussion; Section 4 presents the conclusions.

2. Methods

2.1. The Existing Coordinated TOPSIS

The coordinated TOPSIS proposed by Yu et al. [16] not only takes into account the advantages of TOPSIS, but also can evaluate the coordination level of attributes. However, it should be noted that the TOPSIS adopted in [16] is not the original TOPSIS proposed by Hwang and Yoon in 1981 [1] but modified TOPSIS [19,20]. Therefore, the coordinated TOPSIS proposed by Yu et al. [16] should be called coordinated modified TOPSIS (CM-TOPSIS).

2.1.1. The Principles of CM-TOPSIS

The procedure of CM-TOPSIS consists of the following seven steps [16]:

Step 1: Construct the decision matrix $\mathbf{R} = \{r_{ij}\}$, where r_{ij} is the value of the jth attribute of the ith alternative; $i = 1, 2, \ldots, m; j = 1, 2, \ldots, n$.

Step 2: Normalize the decision matrix $\mathbf{R}$. Two normalization methods are used in CM-TOPSIS [16], not vector normalization (VN) suggested by Hwang and Yoon [1].

For the-bigger-the-better attribute, maximum normalization (MN) is used, which is to divide the values of an attribute by the maximum value of the attribute in all the alternatives. The calculation equation of MN is written as follows:

$$x_{ij} = \frac{r_{ij}}{\max_i r_{ij}}, \tag{1}$$

where x_{ij} is the normalized value of r_{ij}, $\max_i r_{ij}$ is the maximum value of the jth attribute in all the alternatives.

For the-smaller-the-better attribute, min-max normalization (MMN) is used to transform it to be the-bigger-the-better attribute. The calculation equation of MMN is written as follows:

$$x_{ij} = \frac{\max_i r_{ij} - r_{ij}}{\max_i r_{ij} - \min_i r_{ij}}, \tag{2}$$

where $\min_i r_{ij}$ is the minimum value of the jth attribute in all the alternatives. It should be noted that the calculation equation of MMN was written incorrectly in [16].

Step 3: Determine the IS A^+ and the NIS A^-. As MN and MMN are used in Step 2, all the attributes will change to the-bigger-the-better attributes after normalization and the IS A^+ is

$$A^+ = \left\{ \left(\max_i x_{ij} \middle| j \in J \right) \middle| i = 1, 2, \cdots, m \right\} = \{1, 1, \cdots, 1\}. \tag{3}$$

To establish a unique coordination reference standard, an absolute NIS is adopted in CM-TOPSIS [16]. The NIS A^- is

$$A^- = \{0, 0, \cdots, 0\}. \tag{4}$$

Step 4: Calculate the weighted Euclidean distance of each alternative from the IS and the NIS by the following equations:

$$S_i^+ = \sqrt{\sum_{j=1}^{n} w_j(x_{ij}-1)^2}, \quad (5)$$

$$S_i^- = \sqrt{\sum_{j=1}^{n} w_j(x_{ij}-0)^2}, \quad (6)$$

where w_j is the weight of the jth attribute.

Step 5: Calculate the relative closeness (RC) to the IS for each alternative by the following equation:

$$C_i = \frac{S_i^-}{S_i^+ + S_i^-}. \quad (7)$$

Step 6: Calculate the coordination degree (CD) of attributes by the following equations:
For unweighted modified TOPSIS:

$$\cos(\theta_i) = \frac{\sum_{j=1}^{n} x_{ij}}{n\sqrt{\sum_{j=1}^{n} x_{ij}^2}}, \quad (8)$$

$$p_i = \frac{45° - \theta_i}{45°}, \quad (9)$$

where θ_i is the angle between the line from the origin to a point (represents an alternative) and the line from the origin to the IS; p_i is the CD of attributes.

For weighted modified TOPSIS:

$$\cos(\theta_i) = \frac{\sum_{j=1}^{n} w_j x_{ij}}{n\sqrt{\sum_{j=1}^{n} w_j x_{ij}^2}}, \quad (10)$$

$$p_i = \frac{90° - \theta_i}{90°}. \quad (11)$$

Step 7: Calculate the comprehensive evaluation value by the following equation:

$$T_i = (1-v)\frac{C_i}{\max(C_i)} + v\frac{p_i}{\max(p_i)}, \quad (12)$$

where v is the weight of the CD. If the decision or assessment is to encourage the coordinated development of the attributes, v must be greater than or equal to 0.5 [16].

Then, the alternatives can be ranked with respect to their comprehensive evaluation values. A larger comprehensive evaluation value of an alternative indicates that the alternative is relatively better.

2.1.2. The Limitations and Problems of CM-TOPSIS

Although CM-TOPSIS takes into account the coordination level of attributes, which is a significant improvement of TOPSIS, it still has some limitations and problems.

(1) CM-TOPSIS is not based on the original TOPSIS.

The TOPSIS adopted in [16] is not the original TOPSIS proposed by Hwang and Yoon in 1981 [1], but modified TOPSIS [19,20]. In the original TOPSIS, the attribute weights are used to weight the normalization value x_{ij}. In this case, Equations (5) and (6) should be written as follows.

$$S_i^+ = \sqrt{\sum_{j=1}^{n} (w_j x_{ij} - w_j)^2}, \quad (13)$$

$$S_i^- = \sqrt{\sum_{j=1}^{n}(w_j x_{ij} - 0)^2}. \tag{14}$$

Deng et al. [19] used the weighted Euclidean distances instead of the Euclidean distances that are calculated based on the weighted decision matrix. Equations (5) and (6) are the weighted Euclidean distances. The TOPSIS using weighted Euclidean distances is called modified TOPSIS [19,20]. Compared with modified TOPSIS, the original TOPSIS is more frequently used for decision-making and assessment. Therefore, it is significant to establish a coordinated TOPSIS based on the original TOPSIS.

(2) CM-TOPSIS is inapplicable when using VN.

When TOPSIS was proposed, VN was suggested as the normalization method [1], which is frequently used for TOPSIS [21–23]. However, in some studies, VN was replaced by other normalization methods when using TOPSIS, such as MMN [24–26].

An important prerequisite for the application of CM-TOPSIS is that the IS is $\{1, 1, \ldots, 1\}$ and the NIS is $\{0, 0, \ldots, 0\}$. If MMN is used for CM-TOPSIS, this prerequisite can be satisfied. However, if VN is used, the IS is not always $\{1, 1, \cdots, 1\}$, and the NIS is not always $\{0, 0, \cdots, 0\}$. In this case, the prerequisite of CM-TOPSIS cannot be satisfied and the calculate formulas of the CD are inapplicable. Since VN is the most frequently used normalization method for TOPSIS, it should be taken into account in establishing coordinated TOPSIS.

(3) The calculation formulas of the CD are incorrect.

For unweighted modified TOPSIS, taking two-dimensional space as an example, the CD is illustrated in Figure 1. In Figure 1, point N represents an alternative that is described by two attributes. The values of the two attributes are X_N and Y_N, respectively. Point A (1, 1) represents the IS and origin O (0, 0) represents the NIS. OA is the 45° line (coordination line). A point located on the 45° line means that the attributes represented by the point are completely coordinated. The size of ∠NOA indicates the coordination level of attributes. Assuming that point N is located on the 45° line, it can be easily obtained that $X_N = Y_N$. Plugging $X_N = Y_N$ into Equation (8) gives

$$\cos(\theta) = \frac{2X_N}{2\sqrt{2X_N^2}} = \frac{\sqrt{2}}{2}, \tag{15}$$

$$\angle NOA = \theta = 45°. \tag{16}$$

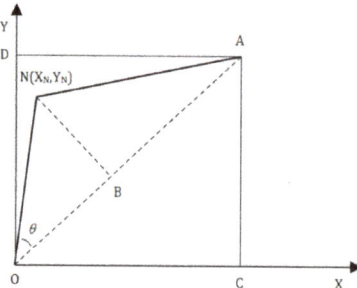

Figure 1. Calculation of the coordination degree (CD) [16].

According to the calculated result of Equation (16), point N should be located on the X or Y axis, which is contradictory to the assumption (N is located on the 45° line). Thus, Equation (8) is incorrect. According to the calculation formula of the angle between two vectors in Euclidean space, the correct calculation formula of $\cos(\theta_i)$ is

$$\cos(\theta_i) = \frac{\sum_{j=1}^{n} x_{ij}}{\sqrt{n}\sqrt{\sum_{j=1}^{n} x_{ij}^2}}. \tag{17}$$

Yu et al. (2018) used $p_i = (45° - \theta_i)/45°$ to indicate the coordination level of attributes [16]. In two-dimensional space, the maximum θ_i is 45°, however, in multi-dimensional space, the maximum θ_i is larger than 45°. In this case, p_i is less than zero, which is irrational. Thus, Equation (9) is incorrect. Equation (11) can be used to replace Equation (9).

For weighted modified TOPSIS, the calculation formula of $\cos(\theta_i)$ is interrelated with the attribute weights, as shown in Equation (10). Assuming that the normalization values of all the attributes are equal for an alternative, Equation (10) can be written as

$$\cos(\theta_i) = \frac{x_{ij}\sum_{j=1}^{n} w_j}{n\sqrt{x_{ij}^2 \sum_{j=1}^{n} w_j}} = \frac{1}{n} \neq 1, n \geq 2. \tag{18}$$

Equation (18) indicates that even if the normalization values of all the attributes are equal, the attributes are not coordinated. This means that the attributes of the alternative represented by the IS $\{1, 1, \ldots, 1\}$ are also not coordinated. For example, a student gets full marks in both Chinese and mathematics. Only considering these two courses, no matter what weight is, it can be obtained that $\cos(\theta) = 1/2$, $\theta = 60°$ and $p_i = 0.333 < 1$. This indicates that the student's learning is very unbalanced in Chinese and mathematics, which is inconsistent with the actual scores. Therefore, the calculation formula of $\cos(\theta_i)$ for weighted modified TOPSIS is incorrect.

Equations (5) and (6) can be written as

$$S_i^+ = \sqrt{\sum_{j=1}^{n} \left(\sqrt{w_j}x_{ij} - \sqrt{w_j}\right)^2}, \tag{19}$$

$$S_i^- = \sqrt{\sum_{j=1}^{n} \left(\sqrt{w_j}x_{ij} - 0\right)^2}. \tag{20}$$

In this case, an equivalent IS is $\{\sqrt{w_1}, \sqrt{w_2}, \cdots, \sqrt{w_n}\}$, an equivalent NIS is $\{0, 0, \ldots, 0\}$, and an equivalent point that represents an alternative is $\{\sqrt{w_1}x_{i1}, \sqrt{w_2}x_{i2}, \cdots, \sqrt{w_n}x_{in}\}$. Obviously, if an equivalent point is located on the line from the equivalent NIS to equivalent IS, the attributes of the alternative represented by the point are completely coordinated. According to the calculation formula of the angle between two vectors in Euclidean space, $\cos(\theta_i)$ can be calculated by the following equation:

$$\cos(\theta_i) = \frac{\sum_{j=1}^{n} w_j x_{ij}}{\sqrt{\sum_{j=1}^{n} w_j} \sqrt{\sum_{j=1}^{n} w_j x_{ij}^2}} = \frac{\sum_{j=1}^{n} w_j x_{ij}}{\sqrt{\sum_{j=1}^{n} w_j x_{ij}^2}}. \tag{21}$$

Assuming that the normalization values of all the attributes are equal for an alternative, Equation (21) can be written as

$$\cos(\theta_i) = \frac{x_{ij}\sum_{j=1}^{n} w_j}{\sqrt{x_{ij}^2 \sum_{j=1}^{n} w_j}} = 1, n \geq 2. \tag{22}$$

Equation (22) indicates that if the normalization values of all the attributes are equal, the attributes are completely coordinated. Thus, Equation (10) should be replaced by Equation (21) when using weighted modified TOPSIS.

If the original TOPSIS is adopted to establish coordinated TOPSIS, Equation (21) should be written as

$$\cos(\theta_i) = \frac{\sum_{j=1}^{n} w_j^2 x_{ij}}{\sqrt{\sum_{j=1}^{n} w_j^2} \sqrt{\sum_{j=1}^{n} w_j^2 x_{ij}^2}}. \tag{23}$$

(4) The coordination level of attributes should be independent of weight.

The weight of an attribute represents the importance of the attribute in decision-making or assessment, but not the importance in the evaluation of coordination level. The coordination level of attributes should be independent of weight, and be evaluated based on the values of attributes. However, Equations (10), (21), and (23) all indicate that the coordination level of attributes is interrelated with the weights, which may cause an irrational result. For example, a student's learning performance is always evaluated based on the scores in the exam, experiment and class discussion. Assuming that there are three students for evaluation, the scores of the three students are listed in Table 1. The weights of the exam, experiment, and class discussion are 0.6, 0.3, and 0.1. An absolute NIS is adopted, which is $\{0, 0, \cdots, 0\}$. Equations (11) and (23) are adopted to calculate the CD. The coefficients of variation (CV) are also calculated for comparison. In terms of the scores, it can be easily judged that the coordination level of Zhang in the three aspects is higher than that of Wang. The CV for the scores of Zhang is 0.088, which is much smaller than that of Wang. However, from Table 1, the CD for the scores of Wang is 0.947, which is larger than that of Zhang. This means that the coordination level of Wang in the three aspects is higher than that of Zhang. A student with a high coordination level of learning should not have a bad performance in class discussion. The contradictory results are caused by the attribute weight. The small weight of 0.1 results in that the score of class discussion has little effect on the coordination level. Even if the score of class discussion is small, a high coordination level can be also obtained. Therefore, the coordination level of attributes should be independent of weight. If the decision-maker believes that the importance of attributes in the evaluation of coordination level is the same as that in decision-making or assessment, the weight should be taken into account in the coordination level. In this case, Equations (21) and (23) can be used.

Table 1. The scores of the three students and calculated results of the CD and coefficients of variation (CV).

Student	Exam Score	Experiment Score	Class Discussion Score	CD	CV
Wang	80	80	35	0.947	0.326
Zhang	70	85	85	0.946	0.088
Li	100	100	100	1.000	0.000

2.2. A Novel Coordinated TOPSIS

In this study, the calculate formulas of the CD for CM-TOPSIS are revised. However, the application of the revised formulas has the same prerequisite as that of the original formulas, which is the IS is $\{1, 1, \ldots, 1\}$ and the NIS is $\{0, 0, \ldots, 0\}$. To avoid this prerequisite and the limitations of CM-TOPSIS, a novel coordinated TOPSIS is proposed in this study.

2.2.1. Original TOPSIS

In this study, the original TOPSIS is adopted to establish the coordinated TOPSIS. The procedure of the original TOPSIS consists of the following six steps [1]:

Step 1: Construct the decision matrix $\mathbf{R} = \{r_{ij}\}$, where r_{ij} is the value of the jth attribute of the ith alternative; $i = 1, 2, \ldots, m$; $j = 1, 2, \ldots, n$.

Step 2: Normalize the decision matrix $\mathbf{R}$ using vector normalization. Other normalization methods can also be used as needed.

Step 3: Calculate the weighted normalized decision matrix $\mathbf{V} = \{v_{ij}\}$ by the following equation:

$$v_{ij} = w_j x_{ij}, \tag{24}$$

where w_j is the weight of the jth attribute; x_{ij} is the normalized value of r_{ij}.

Step 4: Determine the IS A^+ and the NIS A^-:

$$A^+ = \left\{\left(\max_i v_{ij} \middle| j \in J\right), \left(\min_i v_{ij} \middle| j \in J'\right) \middle| i = 1, 2, \cdots, m\right\} = \left\{v_1^+, v_2^+, \cdots, v_n^+\right\}, \quad (25)$$

$$A^- = \left\{\left(\min_i v_{ij} \middle| j \in J\right), \left(\max_i v_{ij} \middle| j \in J'\right) \middle| i = 1, 2, \cdots, m\right\} = \left\{v_1^-, v_2^-, \cdots, v_n^-\right\}, \quad (26)$$

where $J = \{j = 1, 2, \cdots, n | j$ associated with the − bigger − the − better attribute$\}$, $J' = \{j = 1, 2, \cdots, n | j$ associated with the − smaller − the − better attribute$\}$.

Step 5: Calculate the Euclidean distance of each alternative from the IS and the NIS by the following equations:

$$S_i^+ = \sqrt{\sum_{j=1}^n \left(v_{ij} - v_j^+\right)^2}, \quad (27)$$

$$S_i^- = \sqrt{\sum_{j=1}^n \left(v_{ij} - v_j^-\right)^2}. \quad (28)$$

Step 6: Calculate the RC to the IS for each alternative using Equation (7).

2.2.2. Evaluation of the Coordination Level of Attributes

Obviously, if a point is located on the line from the NIS to the IS, the attributes of the alternative represented by the point are completely coordinated. In this case, the following equation can be obtained.

$$\frac{v_{i1} - v_1^-}{v_1^+ - v_1^-} = \frac{v_{i2} - v_2^-}{v_2^+ - v_2^-} = \cdots = \frac{v_{in} - v_n^-}{v_n^+ - v_n^-}. \quad (29)$$

Plugging Equation (24) into Equation (29) gives

$$\frac{x_{i1} - x_1^-}{x_1^+ - x_1^-} = \frac{x_{i2} - x_2^-}{x_2^+ - x_2^-} = \cdots = \frac{x_{in} - x_n^-}{x_n^+ - x_n^-}, \quad (30)$$

where $\{x_1^+, x_2^+, \cdots, x_n^+\}$ and $\{x_1^-, x_2^-, \cdots, x_n^-\}$ are the IS and the NIS without the attribute weights, respectively.

Supposing that $z_{ij} = \frac{x_{ij} - x_j^-}{x_j^+ - x_j^-}$, Equation (30) can be written as

$$z_{i1} = z_{i2} = \cdots = z_{in}. \quad (31)$$

If $x_j^+ = 1$ and $x_j^- = 0$, Equation (31) can be written as

$$x_{i1} = x_{i2} = \cdots = x_{in}. \quad (32)$$

If Equation (31) holds true, the attributes are completely coordinated. Obviously, a large difference among $Z = \{z_{i1}, z_{i2}, \cdots, z_{in}\}$ means a low coordination level of attributes. Therefore, in this study, the diversity of Z is used to indicate the coordination level of attributes, which is independent of weight.

The CV and information entropy (IE) are the two frequently used indicators to measure the diversity of data [27]. In this study, the CV is used to evaluate the coordination level of attributes. The CV can be calculated by the following equation [28]:

$$V_i = \frac{\sigma_i}{\mu_i} = \frac{\sqrt{\frac{1}{n}\sum_{j=1}^n (z_{ij} - \mu_i)^2}}{\mu_i}, \quad (33)$$

where σ_j is the standard deviation of Z; μ_j is the mean value of Z.

The minimum value of the CV is 0, which indicates that the attributes are completely coordinated. The smaller the CV is, the more coordinated the attributes are.

As a comparison, the IE is also used to evaluate the coordination level of attributes. The calculation steps of the IE are described as follows:

Step 1: Normalize $\mathbf{Z}$ by the following equation:

$$F_{ij} = z_{ij} / \sum_{j=1}^{n} z_{ij}. \tag{34}$$

Step 2: Calculate the IE of $\mathbf{Z}$ by the following equation:

$$E_i = -K \sum_{j=1}^{n} F_{ij} \ln F_{ij}. \tag{35}$$

where $K = 1/\ln n$. In particular, when $F_{ij} = 0$, let $\ln F_{ij} = 0$ [29].

The maximum value of the IE is 1, which indicates that the attributes are completely coordinated. The larger the IE is, the more coordinated the attributes are.

In order to combine the RC and CV, the CV needs to be normalized. As the CV is the-smaller-the-better indicator, the following equation is proposed to calculate the comprehensive evaluation value.

$$T_i = (1-v)\frac{C_i}{\max(C_i)} + v\left(1 - \frac{V_i}{\max(V_i)} + \frac{\min(V_i)}{\max(V_i)}\right). \tag{36}$$

If the IE is used, the comprehensive evaluation value can be calculated by the following equation:

$$T_i = (1-v)\frac{C_i}{\max(C_i)} + v\frac{E_i}{\max(E_i)}. \tag{37}$$

Then, the alternatives can be ranked or selected with respect to their comprehensive evaluation values. In this study, the coordinated TOPSIS based on the CV is abbreviated as C-TOPSIS-CV, and the coordinated TOPSIS based on the IE is abbreviated as C-TOPSIS-IE.

Since the coordination level of attributes is independent of weight, Equations (36) and (37) can be also used for modified TOPSIS [19,20] to establish coordinated modified TOPSIS. In this study, the coordinated modified TOPSIS based on the CV is abbreviated as CM-TOPSIS-CV, and the coordinated modified TOPSIS based on the IE is abbreviated as CM-TOPSIS-IE.

3. Results and Discussions

3.1. Case 1: Evaluation of Journals

To illustrate the problem of CM-TOPSIS and compare the CV and IE in evaluating the coordinated level of attributes, the evaluation of JCR2016 robotics journal in [16] is taken as an example. In [16], CM-TOPSIS is used to evaluate the journals. Therefore, in this study, CM-TOPSIS-CV and CM-TOPSIS-IE are adopted to evaluate the journals for comparison. The same indicators, weights, and normalization methods as in [16] are adopted when using CM-TOPSIS-CV and CM-TOPSIS-IE. The absolute NIS is also used, which is $\{0, 0, \cdots, 0\}$. The weight of the CV is set to 0.5, and the weight of the CD or CV or IE is also 0.5. As Equation (10) is incorrect, Equation (21) is also used to calculate the CD. The CM-TOPSIS using the revised calculation formula (Equation (21)) is called revised CM-TOPSIS. The evaluation results of the several methods are listed in Tables 2 and 3. In [16], all the attributes are set to the same weight ($w_j = 1/n$). In this case, Equation (21) can be transformed to Equation (17), which means that the coordinated level determined by revised CM-TOPSIS is independent of weight. Therefore, for this example, the coordinated level determined by revised CM-TOPSIS can be compared with that determined by CM-TOPSIS-CV and CM-TOPSIS-IE.

From Table 2, the calculated results of θ_i and CD have great difference between CM-TOPSIS and revised CM-TOPSIS. For CM-TOPSIS, the CD varies in a narrow domain [0.107, 0.212], which indicates that the coordination level of attributes is low and the difference among the journals is small. However, for revised CM-TOPSIS, the CD varies in a wide domain [0.334, 0.872], which indicates that the difference of the coordination level among the journals is large, and some journals have high coordination level of attributes. Taking the CV as a reference, from Table 3, the CV also has a wide domain [0.203, 1.752], and the largest CV is more than nine times as much as the smallest CV, which also indicates that the difference of the coordination level among the journals is large. Therefore, the calculated results of CM-TOPSIS [16] are irrational, and the revised CM-TOPSIS can be used to replace it.

When the attributes have the same weight or no weight, the CD is independent of weight. Thus, the normalized values of the CD, CV, and IE can be compared. From Tables 2 and 3, the normalized values of the three indicators are different, which may result in different rankings. Figure 2 illustrates the statistical relationships among the normalized CV, normalized IE, and normalized CD. In Figure 2, the 45° line is given. All points on the 45° line have the same values of each component.

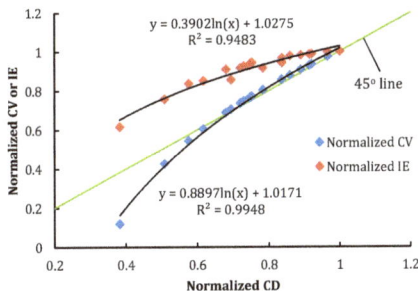

Figure 2. Variation of the normalized CV and information entropy (IE) vs. the normalized CD.

Table 2. Evaluation results of modified Technique for Order Preference by Similarity to Ideal Solution (TOPSIS), coordinated modified (CM)-TOPSIS, and revised CM-TOPSIS.

JCR Abbreviated Title	Modified TOPSIS		CM-TOPSIS [16]				Revised CM-TOPSIS			
	RC	Ranking	θ_i	CD	T_i	Ranking	θ_i	CD	Normalized CD	T_i	Ranking
INT J ROBOT RES	0.610	2	71.479	0.206	0.967	1	17.643	0.804	0.922	0.943	1
SOFT ROBOT	0.633	1	72.970	0.189	0.947	2	28.523	0.683	0.783	0.892	2
IEEE ROBOT AUTOM MAG	0.507	4	72.314	0.197	0.864	3	24.303	0.730	0.837	0.819	3
J FIELD ROBOT	0.426	5	71.539	0.205	0.821	5	18.197	0.798	0.915	0.794	4
BIOINSPIR BIOMIM	0.415	6	71.751	0.203	0.806	6	20.040	0.777	0.891	0.773	5
AUTON ROBOT	0.346	8	70.934	0.212	0.773	8	11.489	0.872	1.000	0.773	6
IEEE T ROBOT	0.515	3	73.132	0.187	0.849	4	35.491	0.606	0.694	0.754	7
ROBOT CIM-INT MANUF	0.401	7	72.060	0.199	0.787	7	22.473	0.750	0.860	0.747	8
ROBOT AUTON SYST	0.302	13	71.140	0.210	0.733	9	14.129	0.843	0.966	0.722	9
SWARM INTELL-US	0.322	12	72.320	0.196	0.718	10	24.344	0.730	0.836	0.672	10
J BIONIC ENG	0.333	9	73.465	0.184	0.696	11	31.371	0.651	0.747	0.636	11
J MECH ROBOT	0.328	11	73.416	0.184	0.694	12	31.103	0.654	0.750	0.634	12
INT J SOC ROBOT	0.329	10	73.853	0.179	0.683	13	33.457	0.628	0.720	0.620	13
J INTELL ROBOT SYST	0.278	15	73.478	0.184	0.652	14	31.444	0.651	0.746	0.593	14
ROBOTICA	0.167	20	72.311	0.197	0.596	15	24.278	0.730	0.837	0.550	15
ADV ROBOTICS	0.193	18	73.395	0.185	0.588	17	30.981	0.656	0.752	0.528	16
INT J ADV ROBOT SYST	0.296	14	76.305	0.152	0.593	16	44.745	0.503	0.576	0.522	17
INT J HUM ROBOT	0.162	21	73.707	0.181	0.555	18	32.685	0.637	0.730	0.493	18
IND ROBOT	0.171	19	74.480	0.172	0.542	19	36.611	0.593	0.680	0.475	19
APPL BIONICS BIOMECH	0.159	22	75.557	0.160	0.504	20	41.560	0.538	0.617	0.434	20
INT J ROBOT AUTOM	0.203	16	77.649	0.137	0.484	21	50.083	0.444	0.508	0.415	21
REV IBEROAM AUTOM IN	0.195	17	80.377	0.107	0.407	22	59.903	0.334	0.383	0.346	22

Table 3. Evaluation results of CM-TOPSIS-IE and CM-TOPSIS-CV.

JCR Abbreviated Title	CM-TOPSIS-IE				CM-TOPSIS-CV			
	IE	Normalized IE	T_i	Ranking	CV	Normalized CV	T_i	Ranking
INT J ROBOT RES	0.975	0.985	0.974	1	0.318	0.933	0.949	1
SOFT ROBOT	0.905	0.914	0.957	2	0.543	0.803	0.901	2
IEEE ROBOT AUTOM MAG	0.955	0.965	0.883	3	0.452	0.856	0.829	3
J FIELD ROBOT	0.974	0.983	0.828	5	0.329	0.927	0.800	4
BIOINSPIR BIOMIM	0.970	0.979	0.817	6	0.365	0.906	0.781	5
AUTON ROBOT	0.990	1.000	0.773	8	0.203	1.000	0.773	6
IEEE T ROBOT	0.846	0.854	0.834	4	0.713	0.705	0.759	7
ROBOT CIM-INT MANUF	0.966	0.976	0.805	7	0.414	0.878	0.756	8
ROBOT AUTON SYST	0.986	0.996	0.737	9	0.252	0.972	0.724	9
SWARM INTELL-US	0.937	0.946	0.728	11	0.452	0.856	0.682	10
J BIONIC ENG	0.919	0.928	0.727	12	0.610	0.764	0.645	11
J MECH ROBOT	0.931	0.940	0.729	10	0.603	0.768	0.643	12
INT J SOC ROBOT	0.905	0.914	0.717	13	0.661	0.735	0.627	13
J INTELL ROBOT SYST	0.926	0.935	0.687	14	0.611	0.763	0.601	14
ROBOTICA	0.931	0.940	0.602	17	0.451	0.856	0.560	15
ADV ROBOTICS	0.932	0.942	0.623	16	0.600	0.770	0.537	16
INT J ADV ROBOT SYST	0.827	0.835	0.651	15	0.991	0.543	0.505	17
INT J HUM ROBOT	0.913	0.922	0.589	18	0.642	0.746	0.501	18
IND ROBOT	0.899	0.907	0.589	19	0.743	0.687	0.479	19
APPL BIONICS BIOMECH	0.842	0.850	0.550	20	0.887	0.604	0.428	20
INT J ROBOT AUTOM	0.748	0.756	0.538	21	1.195	0.425	0.373	21
REV IBEROAM AUTOM IN	0.608	0.614	0.461	22	1.725	0.118	0.213	22

A significant logarithmic function relationship is observed between the normalized IE and normalized CD, and the fitting formula is $y = 0.3902\ln(x) + 1.0275$, the R^2 of which is 0.9483. All the red points are located above the 45° line except point (1, 1), which indicates that the normalized IE is always larger than the normalized CD except the largest value of 1, and the dipartite degree of the normalized IE is lower than that of the normalized CD.

A significant logarithmic function relationship is also observed between the normalized CV and normalized CD, and the fitting formula is $y = 0.8897\ln(x) + 1.0171$, the R^2 of which is 0.9948. Most of the blue points are located close to the 45° line, which indicates that the difference between the normalized CV and normalized CD is small. As shown in Tables 2 and 3, the ranking results determined by CM-TOPSIS and CM-TOPSIS-CV are the same. The blue points located below the 45° line indicates that the normalized CV has a higher dipartite degree when the coordination level of attributes is low.

The Spearman's rank correlation coefficients (SRCC) between the normalized RC and the three normalized indicators are calculated respectively. The calculated results as well as the range of each normalized indicator are listed in Table 4. As shown in Table 4, the range of the normalized IE is 0.386, which is much smaller than that of the normalized CD and normalized CV. This means that the dipartite degree of the normalized IE is lower than that of the normalized CD and normalized CV. Due to the lowest dipartite degree, the SRCC of the normalized IE is the largest, which is 0.935. Since the difference between the normalized CV and normalized CD is small, the SRCCs of the two indicators are the same, and smaller than that of the normalized IE. From Table 3, 77% of the journals have a normalized IE larger than 0.9, which also indicates that the dipartite degree of the normalized IE is low. Therefore, if the decision-maker expects a good dipartite degree of the coordination level, the CD and CV should be used to evaluate the coordination level of attributes. Since the application of the CD has some prerequisites, the CV is recommended.

Table 4. The Spearman's rank correlation coefficients (SRCC) and range of each indicator.

Normalized Indicator	Normalized CD	Normalized IE	Normalized CV
SRCC	0.909	0.935	0.909
Range	0.617	0.386	0.882

3.2. Case 2: Decision-Making for Real Estate Location

Location selection of a real estate project is influenced by many factors. The developers are always concerned about the costs and benefits such as land price and the future development of project. However, for the consumers, they will consider various factors of the real estate project when purchasing a house, such as house prices, traffic conditions, surroundings, and infrastructure. Any drawback of the project may affect the consumers' purchasing desire. Therefore, the coordination level of factors should be taken into account in decision-making for real estate location [18]. To compare C-TOPSIS-IE and C-TOPSIS-CV, and analyze the effect of weight on the coordination level of factors, the decision-making for real estate location [18] is taken as an example. Five factors for decision-making are "educational and medical conditions," "business environment," "land price," "transportation conditions," and "surroundings." The details of each proposed location are listed in Table 5. The 100-point system is used to determine the scores of factors for the five locations [18], as shown in Table 6. The weights of the factors are 0.1, 0.2, 0.3, 0.3, and 0.1, respectively [18].

Table 5. Details of each proposed location [18].

Location	Educational and Medical Conditions	Business Environment	Land Price	Transportation Conditions	Surroundings
Location 1	15 schools and six hospitals nearby. The nearest hospital is 1.6 km away.	One Shopping Mall nearby; six supermarkets within 3km	3.94 million yuan/mu; 1655 yuan/m²	Within the Third Ring Road area; 11 bus lines; Zaoyuan Station of Metro Line 1	Two parks nearby, about 1.5 km away
Location 2	25 schools and eight hospitals nearby. The nearest hospital is 1.9 km away.	Four Shopping Malls nearby; eight supermarkets within 3 km	4.10 million yuan/mu; 1759 yuan/m²	Within the Third Ring Road area; three bus lines; 0.66 km from Baihuacun Station of Metro Line 4 and 2.80 km from Fengcheng Station of Metro Line 2	One park nearby, 3.2 km away
Location 3	23 schools and five hospitals nearby. The nearest hospital is 0.5 km away.	Eight supermarkets within 3 km	4.21 million yuan/mu; 1610 yuan/m²	Within the Second Ring Road area; 18 bus lines; 0.1 km from Xinjiamiao Station of Metro Line 3	Two parks nearby, about 3.2 km away
Location 4	29 schools and five hospitals nearby. The nearest hospital is 1.4 km away.	Two Shopping Malls nearby; 10 supermarkets within 3 km	5.50 million yuan/mu; 1992 yuan/m²	Within the Third Ring Road area; eight bus lines; Jinye Road Station of Metro Line 6 (expected to be completed in 2020)	Five parks nearby; the nearest one is 1.6 km away
Location 5	18 schools and six hospitals nearby. The nearest hospital 2.5 km away.	Central Cultural Business District nearby; five supermarkets within 3 km	3.12 million yuan/mu; 1508 yuan/m²	Outside the Third Ring Road area; five bus lines; Jinsituo Station of Metro Line 4	Four parks nearby; the nearest one is 0.9 km away

Table 6. Scores of the factors for the proposed locations [18].

Factor	Location 1	Location 2	Location 3	Location 4	Location 5
Educational and medical conditions	65	90	90	88	80
Business environment	75	90	75	90	82
Land price	80	85	84	75	95
Transportation conditions	75	65	95	85	80
Surroundings	90	65	85	95	92

In this study, C-TOPSIS-IE and C-TOPSIS-CV are used to determine the optimum location. Since the 100-point system is used, normalization is not required. Absolute IS and NIS are adopted, which are $\{0, 0, \ldots, 0\}$ and $\{100, 100, \ldots, 100\}$ respectively. In this case, Equation (23) can be used to evaluate the coordinated level of factors. Although the NIS is $\{100, 100, \ldots, 100\}$ not $\{1, 1, \ldots, 1\}$, it does not affect the calculation results. The coordinated TOPSIS based on the original TOPSIS and Equation (23) is abbreviated as C-TOPSIS-CD in this paper. The decision results of the several methods are listed in Tables 7 and 8.

Table 7. Decision results of TOPSIS and coordinated (C)-TOPSIS-IE.

Location	TOPSIS		C-TOPSIS-IE		
	RC	Ranking	Normalized IE	T_i	Ranking
Location 1	0.7646	5	0.9983	0.9495	4
Location 2	0.7649	4	0.9949	0.9479	5
Location 3	0.8489	1	0.9998	0.9999	1
Location 4	0.8325	3	0.9998	0.9902	3
Location 5	0.8485	2	1.0000	0.9998	2

Table 8. Decision results of C-TOPSIS-CV and C-TOPSIS-CD.

Location	C-TOPSIS-CV			C-TOPSIS-CD		
	Normalized CV	T_i	Ranking	Normalized CD	T_i	Ranking
Location 1	0.7890	0.8448	4	1.0000	0.9503	4
Location 2	0.5091	0.7050	5	0.9449	0.9230	5
Location 3	0.9781	0.9891	2	0.9823	0.9911	2
Location 4	0.9849	0.9828	3	0.9861	0.9834	3
Location 5	1.0000	0.9998	1	0.9834	0.9914	1

From Table 7, the RCs have little difference between Location 3 and Location 5. Thus, the coordinated level of factors will be an important element affecting the decision result. Due to the low dipartite degree of the IE, the ranking determined by C-TOPSIS-IE is the same as that determined by TOPSIS, and the comprehensive evaluation value have little difference between Location 3 and Location 5. As for C-TOPSIS-CV, the comprehensive evaluation value has a significant difference between Location 3 and Location 5. Location 5 is determined as the best choice, which is the same as that determined by the multi-attribute decision-making method based on balance expectations [18]. From the sales, Location 5 is more popular with consumers [18], which is consistent with the decision result. This validates the feasibility of C-TOPSIS-CV.

The ranking determined by C-TOPSIS-CD is the same as that determined by C-TOPSIS-CV, but the normalized CD is much different from the normalized CV. For example, Location 1 has the largest normalized CD but a normalized CV of 0.7890 that is much smaller than 1.000. This is because that the normalized CD is interrelated with the weights. The three factors of "business environment," "land price," and "transportation conditions" have a total weight of 0.8. If the weight is taken into account, the coordination level will mainly depend on the three factors. As shown in Table 6, the scores of the three factors for Location 1 are 75, 80, and 75, the diversity of which is low. The CV of the three factors for Location 1 is 0.031, which is the smallest among the five Locations. As a result, Location 1 has the largest normalized CD. However, the score of "educational and medical conditions" for Location 1 is just 65, which may severely affect the consumers' purchasing desire. Therefore, the coordination level of factors should be independent of weight so as to obtain a reliable result.

3.3. Discussion

In this study, it was pointed out that the calculation formulas of the CD for CM-TOPSIS are incorrect, and revised calculation formulas of the CD were proposed for CM-TOPSIS. Meanwhile, the calculation formula of the CD for the original TOPSIS was also proposed. However, if the attributes are weighted, these calculation formulas of the CD are interrelated with the weights, which may be result in an irrational result. Therefore, the CV and IE that are independent of the weight were adopted to evaluate the coordination level of attributes. The comparison results show that the dipartite degree of the IE is lower than that of the CV. Therefore, if the decision-maker expects a good dipartite degree of the coordination level, the CV is recommended. The application of the CV has no prerequisite, and it is available for any normalization method. However, the application of the CD has an important prerequisite, which is that the IS is $\{1, 1, \ldots, 1\}$ and the NIS is $\{0, 0, \ldots, 0\}$. If the prerequisite is satisfied or MMN is used, the CD can be used to evaluate the coordination level of attributes.

The CD, IE, and CV are all available for the original TOPSIS and modified TOPSIS to establish coordinated TOPSIS. The calculation formulas of the CD for the two methods are different, while that of the CV or IE is the same for the two methods. However, only when the weights of attributes are equal, the coordination level determined by the CD is independent of weight, and can be compared with that determined by the CV and IE. In this study, it is suggested that the coordination level of attributes should be independent of weight. Therefore, another prerequisite is needed for the application of the CD, which is that the weights of attributes are equal or there is no weight. If the decision-maker believes that the importance of attributes in the evaluation of coordination level is the same as that in decision-making or assessment, the weight should be taken into account in the coordination level. In this case, a weighted CV determined by the following equation can be used.

$$V_i = \frac{\sigma_i^w}{\mu_i} = \frac{\sqrt{\sum_{j=1}^n w_j (z_{ij} - \mu_i)^2}}{\mu_i}. \tag{38}$$

If the weights of attributes are equal, Equation (38) can be transformed to Equation (33). The larger the weight of an attribute is, the greater the influence of the attribute on the coordinated level is.

4. Conclusions

CM-TOPSIS is a significant improvement of TOPSIS, which takes into account the coordination level of attributes in the decision-making or assessment. However, CM-TOPSIS has some limitations and problems listed as follows.

(1) CM-TOPSIS is not based on the original TOPSIS, which is frequently used for decision-making and assessment.
(2) CM-TOPSIS has an important prerequisite, which is that the IS is $\{1, 1, \ldots, 1\}$ and the NIS is $\{0, 0, \ldots, 0\}$. Thus, it is inapplicable when using VN.
(3) The calculation formulas of the CD for CM-TOPSIS are incorrect.
(4) If the attributes are weighted, the coordination level of attributes is interrelated with the weights when using the CD.

In this study, the calculation formulas of the CD for CM-TOPSIS were revised, and that for the original TOPSIS was proposed. The evaluation results of JCR2016 robotics journal indicated that the revision of the calculation formulas of the CD was successful. However, the application of the CD has an important prerequisite, which is that the IS is $\{1, 1, \ldots, 1\}$ and the NIS is $\{0, 0, \ldots, 0\}$. If the prerequisite is satisfied or MMN is used, the CD can be used to evaluate the coordination level of attributes. As the CD is interrelated with the weights, another prerequisite is needed for the application of the CD, which is that the weights of attributes are equal or there is no weight.

To avoid the limitations of CM-TOPSIS, the CV and IE that are independent of the weight were adopted to evaluate the coordination level of attributes. Combining the IE or CV and the original TOPSIS, two types of coordinated TOPSIS called C-TOPSIS-IE and C-TOPSIS-CV were proposed in this study. Two case studies were conducted to compare the IE and CV in evaluating the coordination level of attributes. The comparison results indicate that the dipartite degree of the IE is lower than that of the CV. Therefore, if the decision-maker expects a good dipartite degree of the coordination level, the CV is recommended. Moreover, the CV has no requirement for the normalization method, and is available for the original TOPSIS and modified TOPSIS. It is suggested to use C-TOPSIS-CV or CM-TOPSIS-CV as coordinated TOPSIS, the feasibility of which was validated by the case studies.

Funding: This research was funded by A Project Supported by Scientific Research Fund of Sichuan Provincial Education Department (No. 17ZB0222).

Conflicts of Interest: The author declares no conflict of interest.

References

1. Hwang, C.L.; Yoon, K.P. *Multiple Attribute Decision Making: Methods and Applications*; Springer: New York, NY, USA, 1981.
2. Behzadian, M.; Otaghsara, S.K.; Yazdani, M.; Ignatius, J. A state-of the-art survey of TOPSIS applications. *Expert. Syst. Appl.* **2012**, *39*, 13051–13069. [CrossRef]
3. Zavadskas, E.K.; Mardani, A.; Turskis, Z.; Jusoh, A.; Nor, K.M. Development of TOPSIS method to solve complicated decision-making problems: An overview on developments from 2000 to 2015. *Int. J. Inf. Technol. Decis.* **2016**, *15*, 645–682. [CrossRef]
4. Zyoud, S.H.; Fuchs-Hanusch, D. A bibliometric-based survey on AHP and TOPSIS techniques. *Expert Syst. Appl.* **2017**, *78*, 158–181. [CrossRef]
5. Jozi, S.A.; Shafiee, M.; MoradiMajd, N.; Saffarian, S. An integrated Shannon's Entropy-TOPSIS methodology for environmental risk assessment of Helleh protected area in Iran. *Environ. Monit. Assess.* **2012**, *184*, 6913–6922. [CrossRef]
6. Li, P.; Qian, H.; Wu, J.; Chen, J. Sensitivity analysis of TOPSIS method in water quality assessment: I. Sensitivity to the parameter weights. *Environ. Monit. Assess.* **2013**, *185*, 2453–2461. [CrossRef]
7. Yang, W.C.; Xu, K.; Lian, J.J.; Ma, C.; Bin, L.L. Integrated flood vulnerability assessment approach based on TOPSIS and Shannon entropy methods. *Ecol. Indic.* **2018**, *89*, 269–280. [CrossRef]
8. Nyimbili, P.H.; Erden, T.; Karaman, H. Integration of GIS, AHP and TOPSIS for earthquake hazard analysis. *Nat. Hazards* **2018**, *92*, 1523–1546. [CrossRef]
9. Freeman, J.; Chen, T. Green supplier selection using an AHP-Entropy-TOPSIS framework. *Supply Chain Manag.* **2015**, *20*, 327–340. [CrossRef]
10. Battisti, F.; Guarini, M.R.; Chiovitti, A. The assessment of real estate initiatives to be included in the socially-responsible funds. *Sustainability* **2017**, *9*, 973. [CrossRef]
11. Guarini, M.R.; Battisti, F.; Chiovitti, A. A methodology for the selection of multi-criteria decision analysis methods in real estate and land management processes. *Sustainability* **2018**, *10*, 507. [CrossRef]
12. Mulliner, E.; Malys, N.; Maliene, V. Comparative analysis of MCDM methods for the assessment of sustainable housing affordability. *Omega* **2016**, *59*, 146–156. [CrossRef]
13. Antuchevičienė, J.; Zavadskas, E.K.; Zakarevičius, A. Multiple criteria construction management decisions considering relations between criteria. *Technol. Econ. Dev. Econ.* **2010**, *16*, 109–125. [CrossRef]
14. Yang, W.; Wu, Y. A novel TOPSIS method based on improved grey relational analysis for multiattribute decision-making problem. *Math. Probl. Eng.* **2019**, *2019*, 8761681. [CrossRef]
15. Chen, W. On the problem and elimination of rank reversal in the application of TOPSIS method. *Oper. Res. Manag. Sci.* **2005**, *14*, 39–43.

16. Yu, L.; Yang, W.; Duan, Y.; Long, X. A study on the application of coordinated TOPSIS in evaluation of robotics academic journals. *Math. Probl. Eng.* **2018**, 5456064. [CrossRef]
17. Yu, X.Y.; Liu, H.J. Comprehensive evaluation of information system based on the improved TOPSIS. *Inform. Sci.* **2005**, *23*, 1065–1067.
18. Ma, H.S.; Gao, J.X.; Zhao, Q. Real estate residential location research based on balance expectations. *J. Eng. Manag.* **2016**, *30*, 143–147.
19. Deng, H.; Yeh, C.H.; Willis, R.J. Inter-company comparison using modified TOPSIS with objective weights. *Comput. Oper. Res.* **2000**, *27*, 963–973. [CrossRef]
20. Shyur, H.J. COTS evaluation using modified TOPSIS and ANP. *Appl. Math. Comput.* **2006**, *177*, 251–259. [CrossRef]
21. Karakaş, A.; Kingir, S.; Öztel, A. Evaluation of university employees' work behaviours performance via entropy based TOPSIS methods. *Electron. J. Soc. Sci.* **2016**, *15*, 1046–1058. [CrossRef]
22. Kaynak, S.; Altuntas, S.; Dereli, T. Comparing the innovation performance of EU candidate countries: An entropy-based TOPSIS approach. *Ekonomska Istraživanja* **2017**, *30*, 31–54. [CrossRef]
23. Tiwari, V.; Jain, P.K.; Tandon, P. An integrated Shannon entropy and TOPSIS for product design concept evaluation based on bijective soft set. *J. Intell. Manuf.* **2019**, *30*, 1645–1658. [CrossRef]
24. Ding, L.; Shao, Z.F.; Zhang, H.C.; Xu, C.; Wu, D.W. A comprehensive evaluation of urban sustainable development in China based on the TOPSIS-entropy method. *Sustainability* **2016**, *8*, 746. [CrossRef]
25. Liu, J.; Liu, C.; Han, W. Efficiently evaluating heavy metal urban soil pollution using an improved entropy-method-based TOPSIS model. *Arch. Environ. Con. Toxicol.* **2016**, *71*, 377–382. [CrossRef]
26. Huang, W.; Shuai, B.; Sun, Y.; Wang, Y.; Antwi, E. Using entropy-TOPSIS method to evaluate urban rail transit system operation performance: The China case. *Transp. Res. A* **2018**, *111*, 292–303. [CrossRef]
27. Chen, P. On the diversity-based weighting method for risk assessment and decision-making about natural hazards. *Entropy* **2019**, *21*, 269. [CrossRef]
28. Liu, W.; Li, Q.; Zhao, J. Application on floor water inrush evaluation based on AHP variation coefficient method with GIS. *Geotech. Geol. Eng.* **2018**, *36*, 2799–2808. [CrossRef]
29. Yi, F.; Li, C.; Feng, Y. Two precautions of entropy-weighting model in drought risk assessment. *Nat. Hazards* **2018**, *93*, 339–347. [CrossRef]

© 2019 by the author. Licensee MDPI, Basel, Switzerland. This article is an open access article distributed under the terms and conditions of the Creative Commons Attribution (CC BY) license (http://creativecommons.org/licenses/by/4.0/).

Article

Robust Optimization Model with Shared Uncertain Parameters in Multi-Stage Logistics Production and Inventory Process

Lijun Xu [1], Yijia Zhou [2,*] and Bo Yu [3]

1. School of Science, Dalian Maritime University, Dalian 116026, China; lijun_xu@dlmu.edu.cn
2. School of Computer & Software, Dalian Neusoft University of Information, Dalian 116023, China
3. School of Mathematical Sciences, Dalian University of Technology, Dalian 116024, China; yubo@dlut.edu.cn
* Correspondence: zhouyijia@neusoft.edu.cn

Received: 24 December 2019; Accepted: 2 February 2020 ; Published: 7 February 2020

Abstract: In this paper, we focus on a class of robust optimization problems whose objectives and constraints share the same uncertain parameters. The existing approaches separately address the worst cases of each objective and each constraint, and then reformulate the model by their respective dual forms in their worst cases. These approaches may result in that the value of uncertain parameters in the optimal solution may not be the same one as in the worst case of each constraint, since it is highly improbable to reach their worst cases simultaneously. In terms of being too conservative for this kind of robust model, we propose a new robust optimization model with shared uncertain parameters involving only the worst case of objectives. The proposed model is evaluated for the multi-stage logistics production and inventory process problem. The numerical experiment shows that the proposed robust optimization model can give a valid and reasonable decision in practice.

Keywords: robust optimization; duality theory; uncertain set; logistics production; inventory process

---

1. Introduction

In real life, we usually encounter some problems with uncertain data. In order to solve those problems properly, we must consider them within an uncertain scope. For problems with uncertainty, there are many approaches to handle them, such as sensitivity analysis, stochastic programming, and robust optimization. In this paper, we focus on one of the popular methodologies, robust optimization (RO), which addresses optimization problems with uncertain parameters described using uncertainty sets other than probability distributions. The aim of robust optimization is to determine a solution that is feasible for any realizations of uncertain parameters, and to be optimal for the worst-case scenario of these uncertain parameters. In other words, robust optimization gives a decision which is ensured to be "good" for all possible realizations of uncertain parameters.

1.1. Review of Robust Optimization

In recent decades, robust optimization has been popularized to handle practical problems with incomplete messages, namely, with uncertain data. The primary work by Bertsimas et al. [1] surveys research on robust optimization both theoretically and practically. For linear optimization problems, Brown et al. [2] propose a special robust optimization with constructed uncertainty sets. Ben-Tal et al. [3] relax the standard robustness by varying the protection level across the uncertainty set and extend the framework of robust optimization. There are also some works on robust optimization for the mixed integer stochastic optimization problem. For example, Bertsimas et al. [4] show that there can be a good approximation of the static robust solution for two-stage mixed integer stochastic problem under fairly general assumptions. Delage et al. [5] demonstrate that the associated distributionally robust stochastic programming with a large range of objective functions can be solved efficiently by the proposed model with uncertainty in the form of distribution and moments. Luo et al. [6] investigate robust optimization equilibria in game theory with two players by estimating a bounded asymmetric uncertain set. Xu et al. consider linear regression problems with least-square error in [7]. Later, Xu et al. [8] connect robust optimization and distributionally robust stochastic programming and show that the solution of the RO problem is also a solution of the latter problem. Distributionally robust optimization is later studied as one popular area of robust optimization including convergence, algorithms, and applications, see [9–13] for examples.

In most of those robust optimization models mentioned above, the same uncertain variable does not exist in either objective or constraints simultaneously. However, in some papers like [14–17], the objective and constraints often contain the same uncertain variable simultaneously. In other words, they share the same uncertain parameters. In this case, researchers usually consider the worst cases over the uncertain variable of the objective and each constraint, respectively. For instance, Tong et al. [17] respectively get the dual forms of the worst objective and the worst constraints under the given uncertainty sets. Recently, Yao [18] also employs this method in the robust multi-stage logistics production and inventory process problem. The resulting optimization of this approach is usually a convex and linear problem that is easy to solve. However, these robust models are overly conservative, since the worst case of each objective and each constraint are handled separately. Therefore, they may result in a highly impossible case—that the same uncertain variable is solved with different values in objective and constraints.

1.2. Motivation

Most robust optimization models involve minmax(maxmin) objective functions. The popular approach to handle this kinds of objective is reformulating minmax(maxmin) problems to min(max) ones by the dual theorem. For example, the theoretical work in [11] reformulates the inner maximization as a semi-infinite programming through Lagrange dual when solving minimax distributionally robust optimization problems. Another practical work [10] for economic dispatch in energy integrated systems also convert objective with respect to uncertainty to its dual form. Similarly, for max(min) problems (i.e., the worst cases) in constraints, one strategy is to consider its dual forms with respect to uncertain parameters. Like the work in [17,18] separably transforms the worst-case objective and each constraint to the corresponding dual forms under the given uncertainty sets. However, the separable and respective way to convert the worst-case objective and constraints most likely results in the difference value of the uncertain parameter being between the optimal objective and the worst-case constraint. Recently, Zhou et al. [19] propose robust risk–reward optimization models which ensure the same distribution both in the reward and in constraints. Motivated by the above works, we are prompted to focus on a generalized work which only considers the worst case of the objective itself over the shared uncertain variable instead of also considering the worst case of each constraint. We present an innovative model different to the existing robust optimization models which are overly conservative. In this way, the optimal value of the uncertain parameter in proposed model can be the same one in both the objective and constraints. Moreover, it is less conservative naturally and reasonable in practice.

1.3. Orgnization

In Section 2, we propose the standard form of the robust optimization model with shared uncertain parameters and compare it with the existing robust model. We will show that the proposed robust optimization model is more reasonable. In Section 3, we utilize the approach to remodel a real robust problem in logistics production and inventory process. Then we reformulate them to tractable forms by dual theory under some assumptions. Numerical experiment conducted in Section 4 shows that the existing robust model is too conservative to have a solution in some cases, but the new robust model is validated to forecast earnings and give a good decision to investors. Finally, we conclude this paper in the last section.

2. Robust Optimization Model with Shared Uncertain Parameters

The optimization problems with uncertainty are generally formulated as follows:

$$
\begin{aligned}
\min_{x} \quad & f(x,\xi) \\
\text{s.t.} \quad & g_i(x,\xi) \geq 0, \ i = 1, \cdots, k, \\
& x \in \mathcal{X},
\end{aligned} \tag{1}
$$

where both the objective and constraints are uncertain with the same random variable ξ and the feasible set $\mathcal{X} \subseteq \mathbb{R}^n$. We assume that the variable ξ belongs to a compact uncertainty set U. Generally, the model in Equation (1) is hard to solve because of the existence of uncertain parameter. The popular way to handle this problem, like in [14,16,18,20,21], is to consider the worst case of objective and each constraint, respectively. That is,

$$
\begin{aligned}
\min_{x} \ \max_{\xi \in U} \quad & f(x,\xi) \\
\text{s.t.} \quad & \min_{\xi \in U} g_i(x,\xi) \geq 0, \ i = 1, \cdots, k, \\
& x \in \mathcal{X}.
\end{aligned} \tag{2}
$$

This model is regarded as the classic robust optimization problem.

2.1. Goal and Method

The robust optimization model in Equation (2) considers the worst cases of the objective and all constraints with ξ, respectively. It is straightforward that the uncertain variable ξ in the respective worst case of the objective and constraints are allowed to be different. This may be an overly conservative assumption since it requires that each constraint should be satisfied for all possible (in particular, the worst case) realizations of the uncertain parameters ξ. Therefore, it makes sense to improve these kinds of models, in fact, that one variable should have the same meaning and evaluation at the same scenario. An interesting alternative would be to consider a model in which the optimal uncertain variable in the worst case of the objective itself meets related constraints in Equation (1). Motivated by this consideration, we present the following robust optimization model with shared uncertain parameters (denote as RO_Shared):

$$
\begin{aligned}
\min_{x} \quad & \left\{ \max_{\xi \in U} f(x,\xi) : g_i(x,\xi) \geq 0, \ i = 1, \cdots, k \right\} \\
\text{s.t.} \quad & x \in \mathcal{X}_1,
\end{aligned} \tag{3}
$$

where the feasible set $\mathcal{X}_1 = \{x \in \mathcal{X} : g_i(x,\xi) \geq 0, \xi \in U, i = 1, \cdots, k\}$. We can easily obtain that model in Equation (3) as it only focuses on the worst case of the objective over ξ which indeed satisfies all constraints simultaneously. This model meets the requirement of the real situation more suitably than the model in Equation (2). We illustrate this point by the following example.

2.2. Synthetic Examples and Analysis

Example 1. *Compare the following two robust optimization problems:*

$$
\begin{aligned}
\min_{x \in \mathcal{X}} \ \max_{\xi \in U} \quad & 3x - \xi \\
\text{s.t.} \quad & \min_{\xi \in U} g(x,\xi) = x - \xi^2 \geq 0.
\end{aligned} \tag{4}
$$

$$\min_{x \in \mathcal{X}} \left\{ \begin{array}{ll} \max_{\xi \in U} & 3x - \xi \\ \text{s.t.} & g(x,\xi) = x - \xi^2 \geq 0 \end{array} \right\}. \tag{5}$$

Let $\mathcal{X} = [1,5]$ and $U = [0,2]$. We obtain the optimal solution $x = 4$ and $\xi_1 = 0$ in Equation (4), but the constraint reaches its worst case at $\xi_2 = 2$. So Equation (4) is overly conservative since the value of ξ in the optimal solution may not be the same one as in the worst case of the constraint. If we also consider the worst case of the constraint, we may get different situations which would not likely to happen simultaneously in practice. Meanwhile, the optimal solution of Equation (5) is $(x,\xi) = (1,0)$ which ensures intrinsic consistences of ξ in objective and constraints. Clearly, Equation (5) is less conservative than Equation (4) in terms of optimal objective value.

In fact, the model in Equation (2) is a conservative approximation for the new model in Equation (3). We conclude that the model in Equation (2) is feasible $\Rightarrow$ the model in Equation (3) is feasible.

Denote the non-empty set

$$\mathcal{X}_2 = \left\{ x \in \mathcal{X} : \min_{\xi \in U} g_i(x,\xi) \geq 0, \ i = 1, \cdots, k \right\}$$

as the feasible set of model in Equation (2). According to the definition of the feasible set, for any $x^* \in \mathcal{X}_2$, we have $\min_{\xi \in U} g_i(x^*,\xi) \geq 0, i = 1,...,n$. Obviously, we obtain that $g_i(x^*,\xi) \geq 0, i = 1,...,n$ are satisfied for all $\xi \in U$. Thus x^* is the feasible solution of the model in Equation (3).

The above discussion gives rise to the following elementary result.

Theorem 1. *The feasible set $\mathcal{X}_2$ constitutes a conservative approximation for $\mathcal{X}_1$, that is, $\mathcal{X}_2 \subseteq \mathcal{X}_1$.*

Conversely, if the model in Equation (3) is feasible, the model in Equation (2) is not always feasible, as demonstrated in the following example.

Example 2. *Consider the robust optimization problems in Example 1 with $\mathcal{X} = [1,2]$ and $U = [0,2]$. And then the model in Equation (5) is always feasible for $\forall x \in \mathcal{X}$. Meanwhile, for any $x \in \mathcal{X}$, $x - 4 = \min_{\xi \in U} g(x,\xi) \geq 0$ does not hold, hence the model in Equation (2) is infeasible.*

2.3. Real Example in Portfolio Optimization

Now we take for example the practical portfolio problem. The current robust reward–risk model (see [14–17] for example) is presented as follows:

$$\begin{array}{ll} \min_{x \in \mathcal{X}} & \max_{p \in \mathcal{P}} \text{CVaR}_p(x) \\ \text{s.t.} & \min_{p \in \mathcal{P}} \mathbb{E}_p(x) \geq S_*, \end{array} \tag{6}$$

where $x \in \mathcal{X}$ is the decision variable, $p \in \mathcal{P}$ is an uncertainty set, and S_* represents the lowest expected return. Obviously, Equation (6) is a specific example of Equation (2) by separately considering the worst-case distribution in the reward $\mathbb{E}_p$ and the risk CVaR_p. Intuitively, $\text{argmax}_{p \in \mathcal{P}} \text{CVaR}_p(x)$ may most likely be different from $\text{argmin}_{p \in \mathcal{P}} \mathbb{E}_p(x)$.

However, the corresponding RO_Shared model in Equation (3) for the robust reward–risk model is

$$\min_{x \in \mathcal{X}} \max_{p \in \mathcal{P}} \{ \text{CVaR}_p(x) : \mathbb{E}_p(x) \geq S_* \}. \tag{7}$$

It is straightforward to see the difference on the value of p in the optimal solution of the models in Equations (6) and (7).

It should be clear that the models in Equations (3) and (7) are more flexible than the models in Equations (2) and (6), i.e., it has a larger robust feasible set, enabling a better optimal value while still satisfying all possible realizations of the constraints.

The proposed model in Equation (3) is a generalized model, so we can only provide a framework on how to solve it. That is, the inner maximization problem in Equation (3) can be equivalent to its dual form which is a minimization problem under some assumptions and then the minmax model in Equation (3) can be converted to a minimization problem. Like the specific problems considered in [14–23], we focus on affinely adjustable robust optimization with application to a multi-stage logistics production and inventory process problem.

3. Multi-Stage Logistics Production and Inventory Process

For multi-stage (affinely adjustable) robust optimization such as models in [22,23] and so on, the worst cases of both objective and constraints are also considered and converted to their dual forms, respectively. In this way, they may get different optimal values of the same uncertain variable, which is not true in fact. In this section, we remodel the affinely adjustable robust logistics and inventory problem as the new robust optimization in Equation (3).

3.1. Problem Description

First, we present the following notations.
$\mathfrak{T}$: Set of time intervals $\{1, 2, \cdots, T\}$
I: Set of assets $\{1, 2, \cdots, n\}$
x_i^t: Output of asset i at time t, $i \in I, t \in \mathfrak{T}$
d_i^t: Demand generated in asset i at time t, $i \in I, t \in \mathfrak{T}$
c_i^t: Production cost in asset i at time t, $i \in I, t \in \mathfrak{T}$
p_i^t: Price of asset i at time t, $i \in I, t \in \mathfrak{T}$
P_i^t: Maximum productive capacity of asset i at time t, $i \in I, t \in \mathfrak{T}$
v_i^t: Inventory of asset i at time t, $i \in I, t \in \mathfrak{T}$
m_i^t: Unit inventory cost in asset i at time t, $i \in I, t \in \mathfrak{T}$
Q_i: Maximum productive capacity of asset i, $i \in I$
C: Maximum inventory capacity.

For this problem, the aim is to maximum reward and the deterministic optimization model presented in [18] is as follows:

$$\begin{aligned}
\max_{x,v} \quad & \sum_{i \in I} \sum_{t \in \mathfrak{T}} p_i^t d_i^t - \sum_{i \in I} \sum_{t \in \mathfrak{T}} c_i^t x_i^t - \sum_{i \in I} \sum_{t \in \mathfrak{T}} m_i^t v_i^t \\
\text{s.t.} \quad & v_i^{t+1} \leq v_i^t + x_i^t - d_i^t, \forall i \in I, t \in \{1, \cdots, T-1\}, \\
& 0 \leq x_i^t \leq P_i^t, \forall i \in I, t \in \mathfrak{T}, \\
& 0 \leq \sum_{t \in \mathfrak{T}} x_i^t \leq Q_i, \forall i \in I, t \in \mathfrak{T}, \\
& v_i^t \geq 0, \forall i \in I, t \in \mathfrak{T}, \\
& \sum_{i \in I} v_i^t \leq C, \forall t \in \mathfrak{T}, \\
& v_i^1 = 0, \forall i \in I.
\end{aligned} \qquad (8)$$

Now we consider robust counterpart in terms of uncertain demand d_i^t like in [22]. That is, it is assumed that demand d_i^t is unknown and belongs to a prescribed polyhedral uncertainty set:

$$d_i^t \in U = \{d_i^t : (1-\theta)\hat{d}_i^t \leq d_i^t \leq (1+\theta)\hat{d}_i^t, \sum_{t \in \mathfrak{T}} d_i^t \leq D_i, \forall i \in I, t \in \mathfrak{T}\},$$

where θ is uncertainty level, $\hat{d}_i^t$ is nominal demand in asset i during time interval t and D_i is an upper bound for demand in asset i. The adjustable control variables x_i^t and state variables v_i^t can be represented as the following affine functions of the previously observed demand, i.e.,

$$x_i^t = \eta'_{it} + \sum_{s \in I} \sum_{\tau \in I_t} \eta_{it}^{s\tau} d_s^\tau,$$

$$v_i^t = \pi'_{it} + \sum_{s \in I} \sum_{\tau \in I_t} \pi_{it}^{s\tau} d_s^\tau,$$

where $\eta'_{it}, \eta_{it}^{s\tau}, \pi'_{it}$ and $\pi_{it}^{s\tau}$ are non-adjustable variables and $I_t = \{0, 1, \cdots, t-1\}$.

3.2. The Proposed Model for this Problem

By substituting the state and control variables, we obtain the new affinely adjustable robust counterpart of model in Equation (8) with the shared uncertain d_i^t as follows:

$$\max_{\eta, \eta', \pi, \pi'} \begin{cases} \min_{d \in U} \sum_{i \in I} \sum_{t \in \mathfrak{T}} p_i^t d_i^t - \sum_{i \in I} \sum_{t \in \mathfrak{T}} c_i^t (\eta'_{it} + \sum_{s \in I} \sum_{\tau \in I_t} \eta_{it}^{s\tau} d_s^\tau) \\ \quad - \sum_{i \in I} \sum_{t \in \mathfrak{T}} m_i^t (\pi'_{it} + \sum_{s \in I} \sum_{\tau \in I_t} \pi_{it}^{s\tau} d_s^\tau) \\ \text{s.t.} \quad \pi'_{it+1} + \sum_{s \in I} \sum_{\tau \in I_t} \pi_{it+1}^{s\tau} d_s^\tau \\ \qquad \leq (\pi'_{it} + \sum_{s \in I} \sum_{\tau \in I_t} \pi_{it}^{s\tau} d_s^\tau) + (\eta'_{it} + \sum_{s \in I} \sum_{\tau \in I_t} \eta_{it}^{s\tau} d_s^\tau) - d_i^t, \\ \qquad \forall i \in I, t \in \{1, \cdots, T-1\}, \\ 0 \leq \eta'_{it} + \sum_{s \in I} \sum_{\tau \in I_t} \eta_{it}^{s\tau} d_s^\tau \leq P_i^t, \forall i \in I, t \in \mathfrak{T}, \\ 0 \leq \sum_{t \in \mathfrak{T}} (\eta'_{it} + \sum_{s \in I} \sum_{\tau \in I_t} \eta_{it}^{s\tau} d_s^\tau) \leq Q_i, \forall i \in I, t \in \mathfrak{T}, \\ \pi'_{it} + \sum_{s \in I} \sum_{\tau \in I_t} \pi_{it}^{s\tau} d_s^\tau \geq 0, \forall i \in I, t \in \mathfrak{T}, \\ \sum_{i \in I} \pi'_{it} + \sum_{s \in I} \sum_{\tau \in I_t} \pi_{it}^{s\tau} d_s^\tau \leq C, \forall t \in \mathfrak{T}. \end{cases} \quad (9)$$

s.t. $\pi'_{i1} = 0, \forall i \in I.$

In this model, the objective represents the worst-case reward under the variable d satisfying all related constraints, i.e., the inner problem in Equation (9). We denote all inequality constraints in Equation (9) as $S(\eta, \eta', \pi, \pi', d) \geq 0$ for short. We can utilize duality theory to deal with this complicated max-min problem. We demonstrate this approach in the following theorem.

Theorem 2. *Suppose that, for all fixed $\eta, \eta', \pi,$ and π' concerned in the outer maximizing problem of Equation (9), there exists a $\hat{d} \in U$, such that*

$$S(\eta, \eta', \pi, \pi', \hat{d}) \geq 0.$$

Then the model in Equation (9) is equivalent to the following nonlinear programming problem:

$$
\begin{aligned}
\max_{\lambda,\eta,\eta',\pi,\pi'} \quad & \sum_{i\in I}\sum_{t\in \mathfrak{T}}[-c_i^t\eta_{it}' - m_i^t\pi_{it}' - \lambda_{it}^b\eta_{it}' + \lambda_{it}^c(\eta_{it}' - P_i^t) + \lambda_i^d\eta_i' \\
& -\lambda_{it}^e\pi_{it}' + \lambda_t^f\pi_{it}' + (1-\theta)\lambda_{it}^{ga}\hat{d}_j^t - (1+\theta)\lambda_{it}^{gb}\hat{d}_i^t] \\
& + \sum_{i\in I}\sum_{t\in\{1,\cdots,T-1\}} \lambda_{it}^a(\pi_{it+1}' - \pi_{it}' - \eta_{it}') \\
& - \sum_{i\in I}(\lambda_i^d Q_i + \lambda_i^{gc}D_i) - \sum_{t\in \mathfrak{T}}\lambda_t^f C \\
\text{s.t.} \quad & \sum_{i\in I}\sum_{t\in\{\tau+1,\cdots,T-1\}}\lambda_{it}^a(\pi_{it}^{s\tau} + \eta_{it}^{s\tau} - \pi_{it+1}^{s\tau}) - \lambda_{s\tau}^a + \lambda_{s\tau}^{ga} - \lambda_{s\tau}^{gb} - \lambda_s^{gc} \\
& + \sum_{i\in I}\sum_{t\in\{\tau+1,\cdots,T\}}[(\lambda_{it}^b - \lambda_{it}^c - \lambda_i^d)\eta_{it}^{s\tau} + (\lambda_{it}^e - \lambda_t^f)\pi_{it}^{s\tau}] \\
& + \sum_{i\in I}\lambda_{i\tau}^a\pi_{i\tau+1}^{s\tau} = P_s^\tau - \sum_{i\in I}\sum_{t\in\{\tau+1,\cdots,T\}}(c_i^t\eta_{it}^{s\tau} + m_i^t\pi_{it}^{s\tau}), \\
& \forall s\in I, \tau\in\{1,\cdots,T-2\} \\
& \sum_{i\in I}\pi_{iT}^{s\tau}(-\lambda_{iT-1}^a + \lambda_{iT}^e - \lambda_T^f) + \sum_{i\in I}\eta_{iT}^{s\tau}(\lambda_{iT}^b - \lambda_{iT}^c - \lambda_i^d) \\
& -\lambda_{s\tau}^a + \lambda_{s\tau}^{ga} - \lambda_{s\tau}^{gb} - \lambda_s^{gc} = P_s^\tau - \sum_{i\in I}(c_i^T\eta_{iT}^{s\tau} + m_i^T\pi_{iT}^{s\tau}), \\
& \forall s\in I, \tau = T-1, \\
& P_s^T = \lambda_{sT}^{ga} - \lambda_{sT}^{gb} - \lambda_s^{gc}, \forall s\in I, \\
& S(\eta,\eta',\pi,\pi',\hat{d}) \geq 0, \lambda \geq 0, \pi_{i1}' = 0, \forall i\in I.
\end{aligned}
\qquad (10)
$$

Proof. By the assumption, there must exists a $\hat{d}$ such that $S(\eta,\eta',\pi,\pi',\hat{d}) \geq 0$. Therefore, the inner minimization over d in Equation (9) is feasible. In addition, it is easy to verify the inner minimization in the model in Equation (9) is also bounded below. For each fixed η, η', π and π', we obtain that the strong duality property holds for the inner minimization and then the inner minimization problem can be equivalent to a dual problem (i.e., a maximization problem) by the linear duality theorem. Finally, combining the dual problem with the outer maximization in Equation (9), we can obtain the model in Equation (10). Here we omit detailed derivation since it is an example of the duality theorem in LP. □

Even though we reformulate the affinely adjustable robust optimization as a nonconvex programming problem, many commercial nonlinear optimization solvers can be utilized to solve the problem in Equation (10). Besides, we should note that the model in Equation (10) not only involves less variables but also has more practical significance compared with the model in Equation (12) in [22] and the model in Equation (20) in [18].

4. Evaluation for the Proposed Model

In this section, we apply the proposed robust model of Equation (9) and its equivalent dual form in Equation (10) under the uncertain set U to solve a real multi-stage logistics production and inventory process problem to show the superiority of our robust model. Numerical results will show that our proposed model can give an effective strategy so that investors would receive preferable rewards. More details is as follows.

4.1. Experimental Details

All numerical experiments were run under Matlab version R2015b on a Thinkpad laptop computer with an Intel Core i5 processor at 2.5 GHz with 4 GB RAM. Since the optimization model in Equation (10) is nonconvex and nonlinear, we use the commercial nonlinear optimization solver KNITRO (Available at https://www.artelys.com/en/optimization-tools/knitro) to solve our model.

According to [18], denote logistics products $n = 10$, production planning period $T = 10$, nominal demand $\hat{d}_i^t = 100(1+0.1)^t$, and upper bound for demand in asset i as $D_i = 2\sum d_i^t$, and

$$p_i^t = 100(1+0.05)^t, c_i^t = 30(1+0.05)^t, m_i^t = 2(1+0.05)^t,$$

$$P_i^t = 300(1+0.05)^t, q_i = \sum P_i^t, C = \sum q_i.$$

Next, we compare our model in Equation (10) with the AARC1 model in [18]. The Affinely Adjustable Robust Counterpart (AARC) is first proposed by Ben-Tal et al. in [24]. AARC model restricts the adjustable variables to be affine functions of the uncertain data and converts the NP-hard Adjustable Robust Counterpart (ARC) problem to be a semi-infinite conic programming. Then it can be reformulated as a computationally tractable problem (typically an LP or a semidefinite problem) which is denoted by AARC1 model in [18]. Since the AARC1 model for this multi-stage logistics production and inventory process problem is an LP, we use the commercial solver CPLEX to solve it. Here we test the two models against different levels of uncertainty, specifically, varying θ from 0 to 1 with increment 0.1.

4.2. Results and Evaluations

We output the optimal objective of the two models vs different uncertainty level θ in Figure 1. That is, Figure 1 shows the maximal rewards of two models with different uncertainty levels, respectively. We also plot the histogram of total profit of the 10 periods to comparison more clearly in Figure 2.

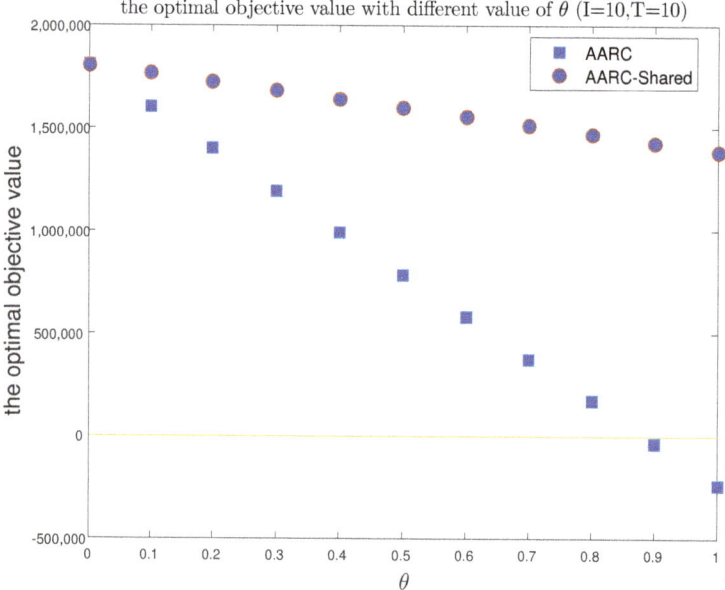

Figure 1. The comparison between our model (Affinely Adjustable Robust Counterpart (AARC)-Shared) and AARC model in [18].

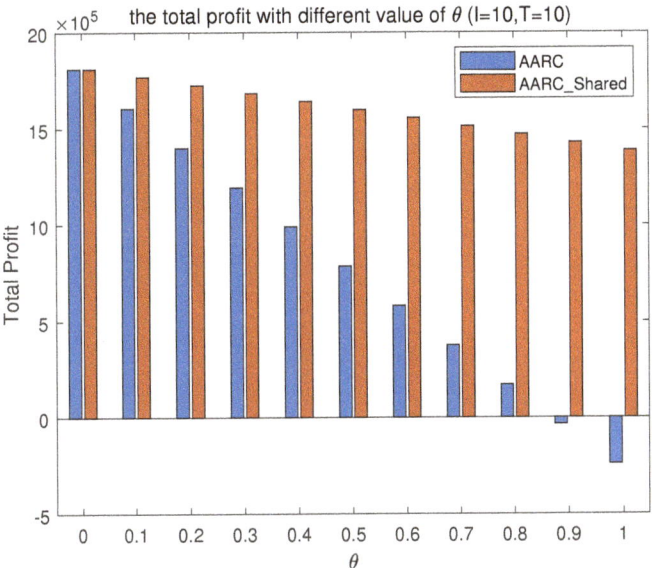

Figure 2. The total profit obtained by two models.

We can see that the maximal rewards decrease along with increasing of the level of uncertainty θ for both models. Here it should be clarified that it is a deterministic multi-stage logistics production and inventory process problem when $\theta = 0$. Obviously, the optimal value of reward in the deterministic case is lager than ones in any other uncertain cases. It exactly illustrates the meaning of robust as we expect. This is easily understandable because the optimal values of both models can be interpreted as the optimistic estimate of total reward in the worst case, which can be lower and lower as the level of uncertainty increases (i.e., the robust feasible region gets larger and larger). This can be interpreted by Figure 3 which shows the demand d_i^5 (the 5th period) with different uncertain level θ.

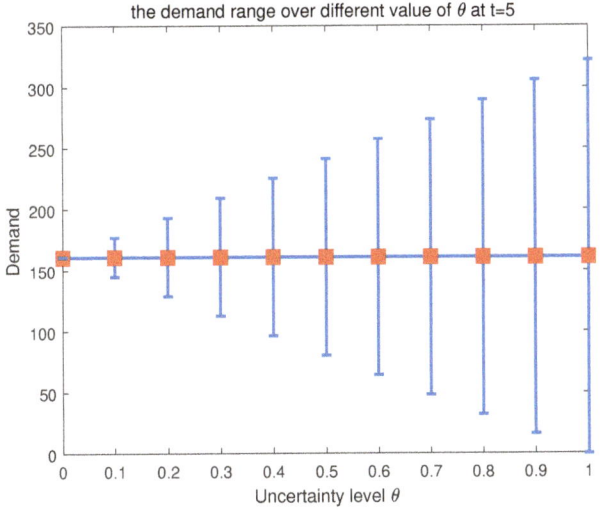

Figure 3. The impact of θ to uncertain demand (e.g., t = 5).

However, compared with our model, the optimal values solved by AARC1 decrease dramatically as θ increasing and specially it can even become negative when $\theta = 0.9$ and 1. This is because AARC1 considers its feasible solutions (for η, η', π, π') satisfying each constraint for all $d_i^t \in U$. Figure 4 plots the range of d_i^t in all 10 periods over $\theta = 0.9$.

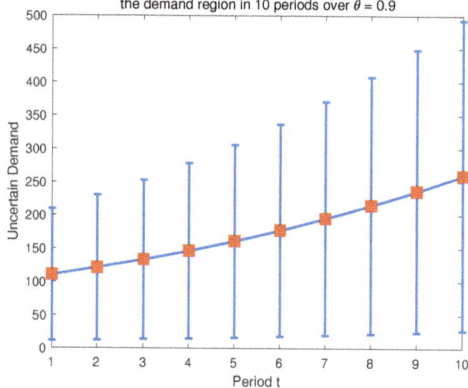

Figure 4. The demand range in all periods over $\theta = 0.9$.

Clearly, as θ increases, the range of d_i^t enlarges which make the outer feasible region of AARC1 shrink rapidly. On the other hand, no matter how the company arranges the production and inventory planning, the model in [18] tells people that they will run at a loss in the worst cases when $\theta = 0.9$ and 1. This is too conservative to give a production and inventory plan. While, our AARC-Shared model is less conservative to some extent and can give a valid strategy in the production and inventory planning under a larger uncertainty level than the AARC1 model.

5. Conclusions

In this paper, the robust optimization model with shared uncertain parameters is proposed. Unlike existing robust models, we only consider the worst case of the objective under the constraints sharing the same uncertain variable and regard it as in the inner problem of entire model. In terms of the duality theorem, the proposed robust model can be reformulated to a nonconvex optimization problem step by step.

Compared with existing robust optimization models in [14,16–18,22–24] and so on, the proposed model has two advantages:

1. Normally, the same variable in one problem indicates the same meaning. So the uncertain situation of the same variable in robust optimization should be the same in one event. In view of this point, our model focuses on the optimization problems under the same uncertain situation shared in both the objective and constraints, and is more practical and realistic.
2. Our model is less conservative and can provide more flexibility in regards to practical problems. When people set more conditions and constraints to models in terms of personal preference, our new robust model can have a greater chance to solve the problems than the original robust models, since they are more conservative with their feasible region. Therefore, the robust optimization model with shared uncertain parameters can be an advisable alternative for investors and the investors can make a decision more actively, but not too conservative by our model.

Author Contributions: Conceptualization, L.X. and Y.Z.; methodology, B.Y.; software, L.X.; validation, L.X., Y.Z., and B.Y.; formal analysis, L.X. and Y.Z.; writing—original draft preparation, Y.Z. and L.X; visualization, L.X.; supervision, B.Y. All authors have read and agreed to the published version of the manuscript.

Funding: This research was funded by Natural Science Foundation of Liaoning Province of China (NO. 2019-BS-013), National Natural Science Foundation of China (NO. 11971092) and the Fundamental Research Funds for the Central Universities (No. DLMU3132019323).

Acknowledgments: We will thank the anonymous reviewers for taking time off their busy schedule to review this paper.

Conflicts of Interest: The authors declare no conflict of interest.

References

1. Bertsimas, D.; Brown, D.B.; Caramanis, C. Theory and applications of robust optimization. *SIAM Rev.* **2011**, *53*, 464–501. [CrossRef]
2. Bertsimas, D.; Brown, D.B. Constructing uncertainty sets for robust linear optimization. *Oper. Res.* **2009**, *57*, 1483–1495. [CrossRef]
3. Ben-Tal, A.; Bertsimas, D.; Brown, D.B. A soft robust model for optimization under ambiguity. *Oper. Res.* **2010**, *58*, 1220–1234. [CrossRef]
4. Bertsimas, D.; Goyal, V. On the power of robust solutions in two-stage stochastic and adaptive optimization problems. *Math. Oper. Res.* **2010**, *35*, 284–305. [CrossRef]
5. Delage, E.; Ye, Y. Distributionally Robust Optimization Under Moment Uncertainty with Application to Data-Driven Problems. *Oper. Res.* **2010**, *58*, 595–612. [CrossRef]
6. Luo, G.M.; Li, D.H. Robust optimization equilibrium with deviation measures. *Pac. J. Optim.* **2009**, *5*, 427–442.
7. Xu, H.; Caramanis, C.; Mannor, S. Robust regression and Lasso. *IEEE Trans. Inform. Theory* **2010**, *56*, 3561–3574. [CrossRef]
8. Xu, H.; Caramanis, C.; Mannor, S. A distributional interpretation of robust optimization. *Math. Oper. Res.* **2012**, *37*, 95–110. [CrossRef]
9. Sun, H.; Xu, H. Convergence analysis for distributionally robust optimization and equilibrium problems. *Math. Oper. Res.* **2016**, *41*, 377–401. [CrossRef]
10. Tong, X.; Sun, H.; Luo, X.; Zheng, Q. Distributionally robust chance constrained optimization for economic dispatch in renewable energy integrated systems. *J. Glob. Optim.* **2018**, *70*, 131–158. [CrossRef]
11. Xu, H.; Liu, Y.; Sun, H. Distributionally robust optimization with matrix moment constraints: Lagrange duality and cutting plane methods. *Math. Program.* **2018**, *169*, 489–529. [CrossRef]
12. Chen, X.; Sun, H.; Xu, H. Discrete approximation of two-stage stochastic and distributionally robust linear complementarity problems. *Math. Program.* **2019**, *177*, 255–289. [CrossRef]
13. Atta Mills, E.; Yu, B.; Zeng, K. Satisfying Bank Capital Requirements: A Robustness Approach in a Modified Roy Safety-First Framework. *Mathematics* **2019**, *7*, 593. [CrossRef]
14. Goldfarb, D.; Iyengar, G. Robust portfolio selection problems. *Math. Oper. Res.* **2003**, *28*, 1–38. [CrossRef]
15. Tütüncü, R.H.; Koenig, M. Robust asset allocation. *Ann. Oper. Res.* **2004**, *132*, 157–187. [CrossRef]
16. Zhu, S.S.; Fukushima, M. Worst-case conditional value-at-risk with application to robust portfolio management. *Oper. Res.* **2009**, *57*, 1155–1168. [CrossRef]
17. Tong, X.J.; Wu, F. Robust reward-risk ratio optimization with application in allocation of generation asset. *Optimization* **2012**, 1–19. [CrossRef]
18. Yao, C. Issues in Multi-stage Logistics Production and Inventory Process Based on Adjustable Robust Optimization. *Logist. Technol.* **2014**, *33*, 231–233.
19. Zhou, Y.; Yang, L.; Xu, L.; Yu, B. Inseparable robust reward–risk optimization models with distribution uncertainty. *Jpn. J. Ind. Appl. Math.* **2016**, *33*, 767–780. [CrossRef]
20. Tong, X.J.; Wu, F.; Qi, L.Q. Worst-case CVaR based portfolio optimization models with applications to scenario planning. *Optim. Methods Softw.* **2009**, *24*, 933–958. [CrossRef]
21. Xu, L.; Yu, B.; Liu, W. The Distributionally Robust Optimization Reformulation for Stochastic Complementarity Problems. *Abstr. Appl. Anal.* **2014**, *469587*, 1–7. [CrossRef]

22. Ben-Tal, A.; Chung, B.D.; Mandala, S.R.; Yao, T. Robust optimization for emergency logistic planning: Risk mitigation in hunmanitarian relief supply chains. *Transp. Res. Part B* **2009**, *45*, 177–189. [CrossRef]
23. Ben-Tal, A.; Golany, B.; Shtern, S. Robust Multi Echelon Multi Period Inventory Control. *Eur. J. Oper. Res.* **2009**, *199*, 922–935.
24. Ben-Tal, A.; Goryashko, A.; Guslitzer, E.; Nemirovski, A. Adjustable robust solutions of uncertain linear programs. *Math. Program.* **2004**, *99*, 351–376. [CrossRef]

© 2020 by the authors. Licensee MDPI, Basel, Switzerland. This article is an open access article distributed under the terms and conditions of the Creative Commons Attribution (CC BY) license (http://creativecommons.org/licenses/by/4.0/).

Article

Integrated Production and Distribution Problem of Perishable Products with a Minimum Total Order Weighted Delivery Time

Ling Liu and Sen Liu *

School of Logistics, Yunnan University of Finance and Economics, Kunming 650221, China; lingliu@ynufe.edu.cn
* Correspondence: liusencool@ynufe.edu.cn

Received: 16 December 2019; Accepted: 19 January 2020; Published: 21 January 2020

Abstract: In this paper, an integrated production and distribution problem for perishable products is presented, which is an NP hard problem where a single machine, multi-customers, and homogenous vehicles with capacity constraints are considered. The objective is to minimize the total order weighted delivery time to measure the customer service level, by making two interacted decisions, production scheduling and vehicle routing, simultaneously. An integrated mathematical model is built, and the validity is measured by the linear programming software CPLEX by solving the small-size instances. An improved large neighborhood search algorithm is designed to address the problem. Firstly, a two-stage algorithm is constructed to generate the initial solution, which determines the order production sequence according to the given vehicle routing. Secondly, several removal/insertion heuristics are applied to enlarge the search space of neighbor solutions. Then, a local search algorithm is designed to improve the neighbor solutions, which further generates more chances to find the optimal solution. For comparison purposes, a genetic algorithm developed in a related problem is employed to solve this problem. The computational results show that the proposed improved large neighborhood search algorithm can provide higher quality solutions than the genetic algorithm.

Keywords: integrated; production scheduling; distribution; large neighborhood search algorithm

1. Introduction

The usage of perishable products has a negative time sensitivity; that is, the usefulness or value of products decreases with time. The definition of perishable products was proposed by [1]. If at least one of the following conditions occurs during the planning period, goods that are raw materials, intermediate products, or final products can be called perishable products: "(1) its physical status deteriorates obvious, and/or (2) its value decreases in the customer's perception, and/or (3) some authorities believe that the future functions may be reduced". Therefore, perishable products have a wide range of definitions, including fresh fruits and vegetables, flowers, food, and other products with a short lifespan, such as blood, drugs, and concrete.

Perishable products should not be delivered long after production, so as to meet customer orders. In addition, companies with such products have employed make-to-order strategies, aiming to reduce the production and delivery lead time [2]. Obviously, for perishable products, it is insufficient to consider the production scheduling or logistics transportation optimization separately. For example, to reduce the value loss of perishable products in delivery, if we make the production start time of the orders as late as possible, it may be difficult to achieve a transportation scheduling of products that is optimal, and the scheduling may be infeasible. In contrast, if only the order transportation optimization is considered, it may lead to excessive production ahead of schedule, leading to a greater value loss. Thus, production scheduling and distribution scheduling decisions should be made jointly at the operational level.

The integrated scheduling of production and distribution has received a lot of attention in recent years [3]. Various problems have been studied for industries such as fashion apparel and toys [4,5], consumer electronics [6], food catering industry [7,8], newspaper [9–11], home healthcare [12], and customized furniture [13]. For perishable products, there is less literature on optimizing production and transportation scheduling simultaneously [2,14–20]. Most of them consider one vehicle to serve all customers, simplifying routing decisions. Therefore, the integrated production and distribution problem for perishable products deserves further study.

In this paper, an integrated production and distribution problem for perishable products (IPDPPP) is presented. As shown in Figure 1, one machine and multi-customer are taken into account. Each customer has only one order, with a different weight according to the importance of each order. The produced products are transported by trucks to customers, without warehousing. In order to make full use of the vehicle capacity, different customers' orders can be loaded into the same vehicle. Obviously, the IPDPPP has to make two decisions—production scheduling and vehicle routing. Production scheduling means to determine the order production sequence. In addition, the orders must be batched first, and the orders in the same batch are produced in succession. Then, each batch is loaded onto a same vehicle. Vehicle routing aims to determine the optimized routing for each vehicle. While the collaborative scheduling problem usually takes customer response time as the first concern, trying to achieve better customer service with lower logistics costs under limited resource constraints [21], the objective of the IPDPPP is to minimize the total order weighted delivery time by making two decisions simultaneously.

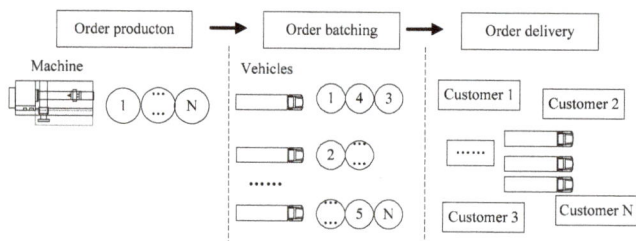

Figure 1. The integrated production and distribution scheduling procedure.

The main highlights are as follows. Firstly, the vehicle routing decision is considered, except for production scheduling, which is simplified in the majority of literature, with respect to the complexity of the problem. Secondly, the objective of this problem is to minimize the total order weighted delivery time, which is usually applied to measure the customer service level [21]. Thirdly, an improved large neighborhood search algorithm is proposed to address the problem. An initial solution is constructed by a two-stage algorithm. Then, the initial solution is improved by exploring a complex neighborhood, and a local search for improving the neighbor solution is developed. Finally, the experimental results show that the improved large neighborhood search algorithm is effective and efficient. Compared with the results of CPLEX and the published genetic algorithm for the related problem, the optimal or approximate optimal solution can be obtained by the improved large neighborhood search algorithm.

The paper is organized as follows. In Section 2, the relevant literature of the IPDPPP is reviewed. In Section 3, a mathematical model for the integrated problem is built. In Section 4, an improved large neighborhood search algorithm is described to solve the problem, and its performance is analysed in Section 5. Section 6 draws the main conclusions.

2. Literature Review

Recently, the IPDPPP has received considerable attention, but there is not much literature about it. In most of the cases, one customer or infinite vehicles are considered, thus vehicle routing is simplified or ignored [22].

For the IPDPPP without routing decisions, an integrated production–distribution model for a deteriorating inventory item was built [23], where one customer was considered. Thus, the routing of the delivery did not need to be decided. The objective was to minimize the total cost. Although considering multi-customers in [24–28], the vehicle routing problem was simplified by direct delivery, which means that each vehicle only served one customer in one trip. The objectives were to minimize the total cost or total transportation time. The IPDPPP considering items with a short lifetime were studied in [29–32]. They assumed that transportation is outsourced to third-party-logistics (3PL) providers, and the products are picked up at regular times. The routing decision was ignored by the manufactories, and only the order production scheduling was optimized. The objectives were to minimize the total profit or total cost. A food production and distribution was mentioned in [33], where infinite vehicles were considered for direct delivery. The objective was to minimize the total cost. A nuclear medicine production and delivery problem considering infinite vehicles was studied in [34]. The objective was to minimize the total cost. The infinite vehicles were also considered in [35], and the objective was to minimize the total cost. A chemotherapy production and delivery problem was addressed in [36]. There was one customer and one vehicle. The vehicle could make more than one trip between the pharmacy production unit and the patient location. The objective to minimize was the maximum tardiness of delivery. Different from the above articles, this paper considers multi-customer and finite vehicles simultaneously; both the order production scheduling and the vehicle routing are optimized. The objective is to minimize the total order weighted delivery time in this paper, which differs from these papers.

The IPDPPP becomes more complex when routing decisions are considered. The IPDPPP with time windows was addressed in [16]. They assumed that the demands of customers are stochastic. The aim was to maximize the expected total profit of the supplier. The IPDPPP considering the time windows and parallel production lines were studied in [18,37], the goals were to minimize the total cost. An integrated supply chain scheduling problem along with the batch delivery consideration was investigated in a series of multi-factory environments [38], while considering the due date of each order. The aims were to reduce the total cost of transportation and the total tardiness. The products with a short lifespan were considered in [2,14,15,17,19,20], as well as the order production sequence and delivery sequence were fixed. A truck that has enough capacity to deliver all of the orders in one trip was mentioned in [14,17]. The goal was to maximize the total demand without violating the production/distribution capacity, the product lifespan, and the delivery time window, by selecting a subset of customers from a given sequence to receive the deliveries. While one truck with a limited capacity for transportation was considered in [2,15], which could travel many trips. The goal was to minimize the maximum delivery time of the orders. Multi-trucks with a limited capacity for transportation was addressed in [19,20], which could travel many trips. Their objectives were to minimize the total cost and makespan, respectively. Comparatively, we relaxed the time windows constraint in this paper, which further increases the complexity of the problem due to an expansion in search space. The time windows constraint was also relaxed in [22], the scheduling of production and delivery were considered in a make-to-order environment considering a single machine, multi-customers, and multi-vehicles. The objective was to minimize the makespan and an improved genetic algorithm was proposed. In this paper, however, the objective is to minimize the total order weighted delivery time, which receives relatively less attention in existing literature.

As a result of the complexity in integrated production scheduling and transportation problems, they were solved, in general, via heuristic algorithms. Such as, the branch-and-bound algorithm [2,14], the genetic algorithm [15], the Nelder–Mead method with a heuristic algorithm [16], a new heuristic algorithm [17], the adaptive large neighborhood search [18], a heuristics algorithm based on evolutionary algorithms [19], a greedy randomized adaptive search procedure with an evolutionary local search [20], a Benders decomposition-based heuristic [36], a novel three-phase methodology [37], and a Pareto approach [38]. What is more, in other areas of combinatorial optimization, new algorithms have emerged, such as multi-criteria optimization [39], and a biased-randomized iterated local search algorithm [40]. In this paper, an improved large neighborhood search algorithm is considered. By applying the proposed algorithm, the neighbors could be enlarged by their removal heuristics and insertion heuristics. Then, the solution space is enlarged, which generates more chances to find the optimal solution.

3. Problem and Model Definition

In this section, a mathematical model of the IPDPPP is developed. In the production phase, one machine is considered to produce products from all of the orders at a plant, with a constant production rate of η. The machine will not be able to produce a new order until the order being processed has been completed. $P = \{0, 1, 2, \ldots, n\}$ denotes the set of the plant and all of the customers, with 0 representing the plant and $N = \{1, 2, \ldots, n\}$ representing the customers. $A = \{(i, j); i, j \in P\}$ denotes the set of edges, with each edge having a travelling time of b_{ij} (from the plant to a customer or from a customer to another customer).

Customers are in different geographical locations. Each customer places an order to the plant, and $F = \{f_1, f_2, \ldots, f_n\}$ is the set of all orders. w_i denotes the weight of the order, f_i, which presents the importance of the order, f_i. Each order, f_i, has a definite demand, d_i, then the order processing time, t_i, can be calculated according to the formula $t_i = d_i / \eta$.

$U = \{1, \ldots, H\}$ is the set of multiple homogenous vehicles with a capacity of Q. Each vehicle can be used at the most once, starting and ending at the plant. The loaded order quantity of a vehicle cannot exceed the vehicle's capacity. Each customer's demands must be delivered at one time, not in batches. Orders belonging to different customers can be delivered by the same vehicle within one trip. We defined the departure time as the time when the vehicle leaves the plant. The objective is to minimize the total weighted delivery time of the orders, which is clearly influenced by the departure time and the transportation time of each vehicle.

The other notations are defined as follows:

Decision variables:

$$x_{ij}^h = \begin{cases} 1, & \text{if vehicle } h \text{ visit the edge } (i,j) \\ 0, & \text{otherwise} \end{cases} \quad y_i^h = \begin{cases} 1, & \text{if vehicle } h \text{ is loaded with order } f_i \\ 0, & \text{otherwise} \end{cases}$$

$$z_{ij} = \begin{cases} 1, & \text{if order } f_i \text{ is produced prior to order } f_j \\ 0, & \text{otherwise} \end{cases}$$

Variables:

V_i: the production completion time of order f_i D_i^h: the arrival time of vehicle h at customer i

Objective function:

$$Min \sum_{h=1}^{H} \sum_{j=1}^{n} w_j D_j^h \tag{1}$$

Constraints:

$$\sum_{h=1}^{H} \sum_{i=0}^{n} x_{ij}^h = 1 \quad j = 1, 2, \ldots, n \tag{2}$$

$$\sum_{i=1}^{n} x_{i0}^h = 1 \quad h = 1, 2, \ldots, H \tag{3}$$

$$\sum_{j=1}^{n} x_{0j}^h = 1 \quad h = 1, 2, \ldots, H \tag{4}$$

$$\sum_{i=0}^{n} x_{ig}^h - \sum_{j=1}^{n} x_{gj}^h = 0 \ g = 1, 2, \ldots, n, \ h = 1, 2, \ldots, H \tag{5}$$

$$\sum_{i=0}^{n} \sum_{j=1}^{n} x_{ij}^h d_j <= Q \ h = 1, 2, \ldots, H \tag{6}$$

$$C_0 = 0 \tag{7}$$

$$V_i + t_j - V_j \le (1 - z_{ij})M \ i = 0, 1, 2, \ldots, n, \ j = 1, 2, \ldots n \tag{8}$$

$$\sum_{i=0}^{n} z_{ij} = 1 \ j = 1, 2, \ldots, n \tag{9}$$

$$\sum_{j=1}^{n+1} z_{ij} = 1 \ i = 1, 2, \ldots, n \tag{10}$$

$$D_i^h + b_{ij} - D_j^h \le (1 - x_{ij}^h)M \ i = 0, 1, 2, \ldots, n, \ j = 1, 2, \ldots, n, \ h = 1, 2, \ldots, H \tag{11}$$

$$D_i^h \ge 0, \ i = 0, 1, 2, \ldots, n, \ h = 1, 2, \ldots, H \tag{12}$$

$$D_0^h \ge \max_{j \in N}(V_j y_j^h), \ h = 1, 2, \ldots, H \tag{13}$$

$$y_j^h = \sum_{i}^{n} x_{ij}^h \ j = 1, 2, \ldots, n, \ h = 1, 2, \ldots, H \tag{14}$$

$$x_{ij}^h \in \{0, 1\} \ y_j^h \in \{0, 1\} \ z_{ij} \in \{0, 1\} \ i = 0, 1, 2, \ldots, n, \ j = 1, 2, \ldots, n, \ h = 1, 2, \ldots, H \tag{15}$$

Objection Function (1) aims at minimizing the total weighted delivery time of the orders. Constraint (2) indicates that each customer's demands must be met at one time. Constraints (3) and (4) ensure that each vehicle can be used once, starting and ending at the plant. Constraint (5) represents that the entering vehicle of a customer must eventually leave that customer. Constraint (6) shows that the vehicle capacity cannot be exceeded. Constraint (7) represents the starting time of a machine. Constraints (8) to (10) ensure that each order has a single predecessor and a single successor in the production phase. Constraints (11) to (12) ensure that each customer has a single predecessor and a single successor in the delivery phase. Constraint (13) ensures that the departure time of each vehicle is greater than or equal to the latest production completion time among the orders onboard the vehicle. Constraint (14) indicates that each customer has to be served. Constraint (15) is the integer conditions.

The IPDPPP is an NP-hard problem, optimizing both production scheduling and vehicle routing [22]. Intelligent algorithms have made great achievements in solving production scheduling [41,42], and vehicle routing problems [43,44]. For the complexity of the integration of the production scheduling and distribution problem, metaheuristic can be an appropriate approach to solve it [38]. Therefore, a heuristic algorithm is developed.

4. An Improved Large Neighborhood Search Algorithm

The specialty of our problem is that production scheduling and vehicle routing are considered jointly, which interplay with each other. Production scheduling determines the production sequence of every order; the vehicle routing problem determines the delivery sequence of each order for every vehicle. It is remarkable that the vehicle routing problem is a weighted traveling repairman problem (TRP), not a traditional traveling salesman problem (TSP), as the aim of this research is to minimize the total order weighted delivery time in this research. The difference between them is that the TRP aims at minimizing the sum of all the customers' delivery times by taking into account customers' satisfaction [45,46]; while the TSP aims at minimizing the total distances. For example, for a TRP problem or a weighted TRP, if the sequence of all of the customers in a route is reversed, the objective value would be changed; however, for a TSP problem, the objective value is not affected. Therefore, the heuristics to discuss the integration of production and transportation in the literature, with the goal of total cost or makespan, is not suitable to solve this problem.

With respect to the above special consideration, an improved large neighborhood search (ILNS) algorithm is proposed. The large neighborhood search (LNS) algorithm is a meta-heuristic algorithm, which finds better candidate solutions by exploring complex neighborhoods defined by destroy and repair methods. A destroy method destructs part of the current solution, while a repair method rebuilds the destroyed solution. By alternating between an infeasible solution and a feasible solution through destroy and repair methods, the LNS algorithm has been widely used in various combinatorial optimization problems, and has been proved effectively, such as for vehicle routing problems [47–49], machine scheduling problems [50,51], supply chain network design [52], satellites scheduling [53], and exam scheduling problems [54]. By applying the LNS algorithm to solve our problem, the neighbors could be enlarged, so as to solve the solution space. Compared with the traditional LNS algorithm, in addition to the above benefits, the improved large neighborhood search (ILNS) algorithm proposed in this paper uses a two-stage algorithm to construct the initial solution, and proposes a local search to improve the neighbor solution, which generates more chances to find the optimal solution. The ILNS algorithm includes the following five stages, and the major procedure of the ILNS algorithm is shown in Figure 2.

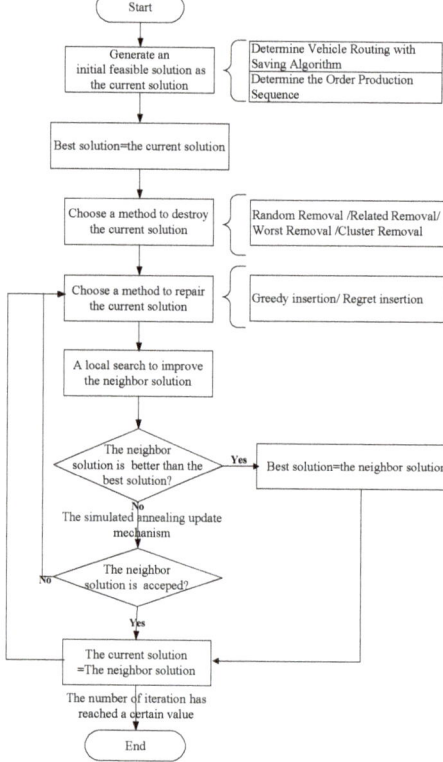

Figure 2. The major procedure of the improved large neighborhood search algorithm.

Stage 1: Initial solution generation. A two-stage algorithm is developed in the generation. The vehicle routing is determined with the savings algorithm first. Then, the order production sequence is determined by a certain rule, and the objective value of the solution is computed. The constructed initial solution is used as the current solution. The best solution is also equal to the initial solution at this stage.

Stage 2: Neighbor solution generation. A neighborhood is defined implicitly by a destroy and a repair method. In this paper, the destroy method consists of four removals, and the repair method consist of two insertion algorithms. Every time a neighbor solution is derived with the current solution, a certain removal and insertion algorithm is chosen with a certain rule. Each time the vehicle routing is generated, the order production sequence is determined, and the objective value of the neighbor solution is computed.

Stage 3: Local search. A local optimization algorithm is employed to improve the quality of the generated neighbor solutions.

Stage 4: Acceptance rule. If the newly derived neighbor solution is better than the best solution, we update the best solution and set it to be the new current solution; otherwise, the judgement criterion of the simulated annealing algorithm will be used to decide whether to accept the neighbor solution as the new current solution or not.

Stage 5: Stopping criterion. Such an iteration process (Stages 2 and 4) will be repeated until the defined number is reached.

4.1. Construction of an Initial Solution

A two-stage algorithm is proposed to construct the initial solution. First, the vehicle routing is determined with a savings algorithm. When the vehicle routing is given, the orders loaded on each vehicle are called a batch. Second, the production sequence of the orders in each batch is generated based on the vehicle routing. Finally, the objective value is calculated.

4.1.1. Determine Vehicle Routing with a Saving Algorithm

The saving algorithm for the vehicle routing problem is introduced in [55]. For the initial solution, each customer is assigned to a separate route. Then, for each customer i in route λ_1 and j in route λ_2, the savings are calculated as follows: $s_{ij} = b_{i0} + b_{0j} - b_{ij}$; 0 represents the depot and b_{ij} represents the cost of edge (i, j). Thus, the value of s_{ij} contains the savings of combining two routes λ_1 and λ_2, instead of serving two separate routes. A pseudo-code for the saving algorithm is presented in Algorithm 1.

Algorithm 1: Saving Algorithm

Set $X = 1$, $C = \{1,2, \ldots, n\}$, $S = \{s_{ij}: i,j \in C\}$
Insert n customers into n empty routes.
Calculate the savings s_{ij} between any two customers
Sequence s_{ij} in S in non-increasing order
While S is not empty do
Mark the largest savings s_{ij}
If the onboard quantity of vehicle X does not exceed its capacity when i and j are loaded, then
Append the arc (i,j) to the end of route X
Remove arc (i,j) and other arcs that contain point i or j from set S
else
$X = X + 1$
For each customer c in C, do
If customer c is not loaded to any route, then
Load the order of customer c into the route that has the largest remaining capacity
Return to the route of each vehicle

4.1.2. Determine the Order Production Sequence

As the vehicle routing decision has been made, the transportation time and the orders loaded on each vehicle are determined. The departure time of each vehicle should be reduced, because the objective is to minimize the total weighted delivery time of the orders. The departure time of a vehicle depends on the order production sequence, which is determined as follows.

The rule that the orders onboard the vehicle with the maximum sum of order weighted delivery times are produced first is employed. Assume that the departure times of all of the vehicles are 0. Then, calculate the sum of order weighted delivery times of each vehicle. Finally, arrange the production sequence of each vehicle as per the sum of the order delivery times of all vehicles, sorted in descending order. Assume that the orders that belong to the same order batch are produced successively, and that the production sequence of each order is the same as its delivery sequence; re-calculate the departure time of every vehicle and the sum of order weighted delivery times of each vehicle. Then, the total weighted delivery time of the orders is obtained. The pseudocode to determine the order production sequence is presented in Algorithm 2.

Algorithm 2: Determine the Order Production Sequence

(1) Set the departure time L_λ of each route λ to 0, $\lambda \subset \{1, 2, \ldots, H\}$;
(2) Calculate the sum of the order weighted delivery times S_λ of each route λ;
(3) Sort all routes in descending order of S_λ;
(4) Produce orders of each route in above turn;
(5) Re-calculate each L_λ and S_λ;
(6) $objective = sum(S_\lambda)$;

Return *objective*

4.2. Neighborhood Search

Given a current solution, s, several customers are removed and are then reinserted to generate neighbor solutions at each iteration. This is achieved by applying one of several removal and insertion heuristics. Each time a neighbor solution is generated, the objective value of the neighbor solution is calculated as the initial solution.

4.2.1. Four Removal Heuristics

The removal heuristic is to generate destroy neighborhoods by choosing m customers from a current solution, s, and putting them into a request bank, such as random removal, related removal, worst removal, and cluster removal, which can be seen in Figure 3. The black circles denote the removed customers and the dashed lines denote the new generated edges after moving (Figure 3).

(1) Random removal

The simplest removal heuristic randomly selects m customers and deletes them from the current solution, s, which is propitious to diversify the search.

(2) Related removal

The general idea of the related removal heuristic aims to delete somewhat similar customers, as it is considered fairly easy to create new and better solutions [47]. The similarity between two customers, i and j, is calculated by a correlation measure $R_{ij} = d_{ij}$ (the distance between two customers), where the lower value means the more similar customers. The related removal heuristic repeatedly chooses a new customer, i, randomly, and the customer i^r, having the smallest relatedness with i, removes them from the current solution, s, until m customers have been removed.

(3) Worst removal

This heuristic approach eliminates the customers with a high-cost in the current solution s. Let $Cost(s,i)^- = f(s,i) - f(s,i^-)$; $f(s,i)$ is the cost associated with customer i in the current solution s, and $f(s,i^-)$ is the cost without customer i in s. The worst removal heuristic repeatedly selects a new customer i, with the highest $Cost(s,i)^-$, until m customers are deleted [56].

(4) Cluster removal

Some instances tested in the computational section contain clusters of customers; it needs to remove clusters of related customers from some routes. For each route, a modified version of Kruskal's algorithm is applied to divide its customers into two groups, it is stopped when two connected components are left [56]. Then, one of the groups is selected randomly and the customers of the group are removed so that they can be re-inserted appropriately. If more customers need to be removed, one of the removed customers is selected, and a customer is picked from a new route λ_{new} that is closer to the chosen customer. The new route, λ_{new}, is then divided into two clusters, and the process is repeated until m customers are deleted.

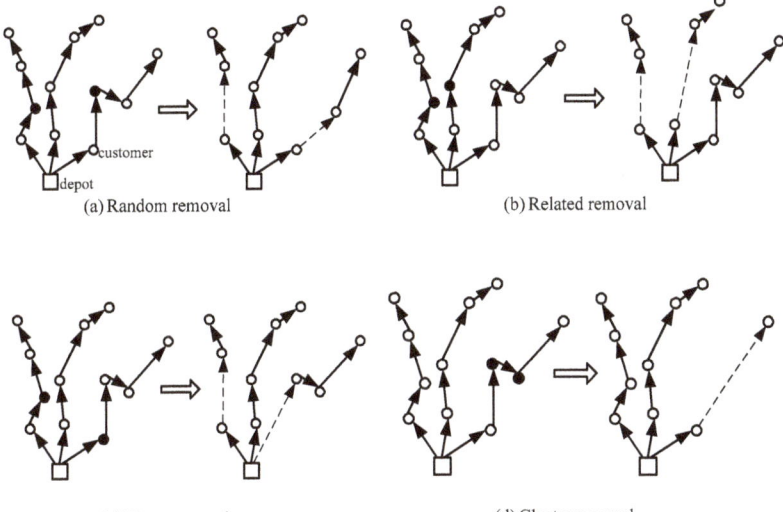

Figure 3. Four removal heuristics.

4.2.2. Two Insertion Heuristics

The insertion heuristic is used to rebuild the destroyed solution. When the removal heuristics remove customers from the existing routes into a request bank, to generate repair neighborhoods, the insertion heuristic choose m customers from the request bank, inserting them into one or more routes without violating the capacity constraint.

(1) Greedy insertion

Insert customer i into route k in the position that results in the lowest objective value. This process continues until m customers have been inserted. If customer i cannot be reinserted for the capacity constraint, leave it in the request bank. Finally, insert the customers remaining in the request bank into a route randomly, and let the objective value of the new solution be a very large number.

(2) Regret insertion

For each removed customer in the request bank, calculate its regret value, this is equal to the difference in cost between two solutions, in which i is inserted in its best route or in its second-best route [47].

Let $\Delta f(i^1)$ denote the change in the objective value by inserting customer i in the route λ_1, where customer i can be inserted at a minimum cost. Let $\Delta f(i^2)$ denote the change in the objective value by inserting customer i in route λ_2, where the customer i can be inserted at the second minimum cost. The regret value of customer i is equal to $\Delta f(i^2) - \Delta f(i^1)$.

Customer i with the highest regret value is chosen for insertion into the current solution, s, and customer i is deleted from the request bank. Customer i is inserted in its best route. The process is repeated until the request bank is empty.

4.3. A Local Search for Improving the Neighbor Solution

A multiple insertion (MI) algorithm was adapted for parallel machines scheduling [57]. The MI heuristic sorts the jobs in a non-increasing order of the modified processing times, then places each job in the position on the machine with the lowest makespan. In this paper, the MI algorithm is adopted to improve the quality of the neighbor solution. The vehicles can be treated as parallel machines, and the travelling time between customers can be the modified processing times. Thus, sort the customers in a non-increasing order of the travelling time, insert each customer one by one in every position of every route, and then place each customer in the position that results in the lowest objective value.

4.4. Acceptance Rule

The current solution update mechanism determines whether to replace the current solution s by the neighbor solution s'. The acceptance rule of the simulated annealing algorithm is used as the mechanism to accept neighbor solutions [58].

There are two situations that need to be considered. One is that if neighbor solution s' is superior to the current solution s—replace s with s'. Another is that if neighbor solution s' is inferior to the current solution s, the acceptance probability $\exp \frac{(v(s')-v(s))}{T}$ needs to be computed, with that, $v(s)$ and $v(s')$ are the objective values of the current solution and the neighbor solution, respectively. T is the current temperature, which starts at 0.005× (the objective value of the initial solution). At every iteration, similar to [47], T decreases linearly to zero, according to a cooling rate fixed to 0.99975. If the acceptance probability is larger than a random number between $[0, 1]$, accept the neighbor solution s' and continue searching in the current neighborhood structure; otherwise, turn to the next neighborhood structure to search.

4.5. Stopping Criterion

The removal and insertion heuristics are repeated until the number of iterations reaches 50,000 [47].

5. Computational Results

5.1. Instances Generation

There is no benchmark data for the IPDPPP, the test instances for the computational experiments should be generated according to certain rules. The first set of instances consists of small-sized instances including 5–10 customers and two vehicles. The vehicles are homogenous with a capacity of 20. The coordinates of each customer are randomly generated in [0, 50]. The demand of each customer/order is randomly generated in [1, 10], and the total demands of all of the customers does not exceed 40.

If the order production time is ignored, the problem IPDPPP is simplified to a capacitated vehicle routing problem (CVRP). Thus, the second set of instances consists of larger-sized instances generated from the well-known CVRP benchmarks (A, B, and P series), including 71 instances found on the website http://www.bernabe.dorronsoro.es/vrp/. The instances are composed of 15–100 customers and 2–15 vehicles. The number of customers, customer coordinates, and customers'/orders' requirements in the set of data are the same as the CVRP benchmark data. The characteristics of the customer coordinates of groups A, B, and P are different. Take the instances A-n32-k5, B-n31-k5, and P-n40-k5 as examples; the customer coordinates of Group A are generated randomly, while that of Group B are clustered and the customer coordinates of Froup C have an almost equal distance, which can be seen in Figure 4.

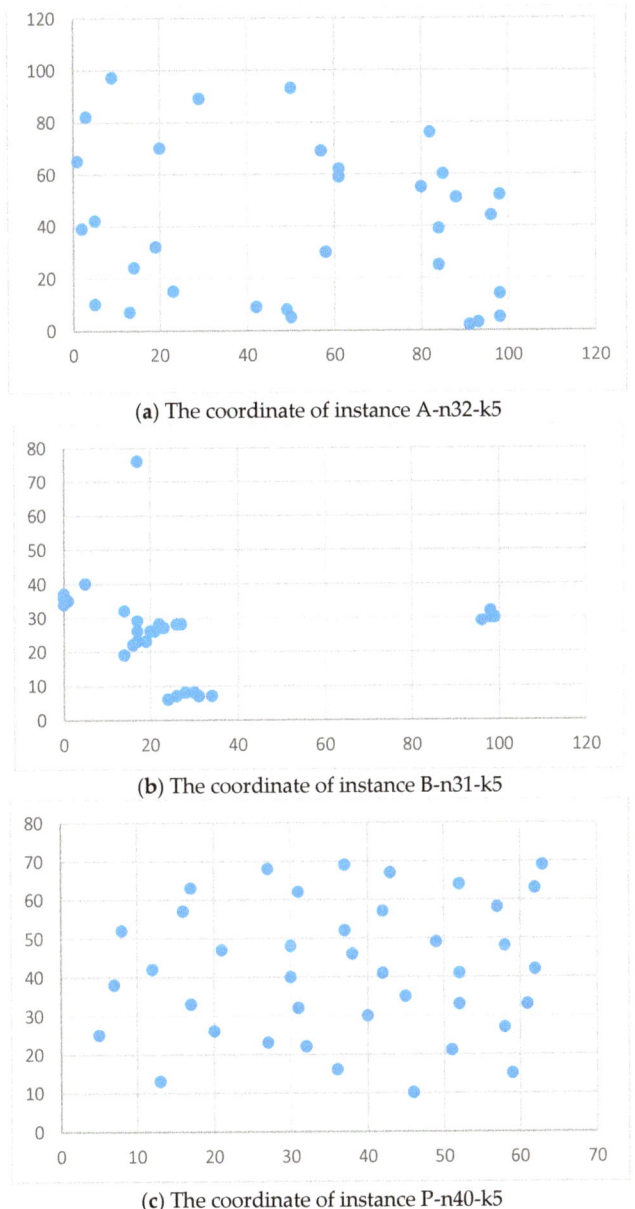

(a) The coordinate of instance A-n32-k5

(b) The coordinate of instance B-n31-k5

(c) The coordinate of instance P-n40-k5

Figure 4. Examples of the coordinates of instances in Groups A, B, and P.

For each instance, both in first and second set, the order weight is selected from the discrete uniform distribution U [1, 5], at random. For ease of calculation, similar to [15], the production rate is set to 1. In addition, similar to [44], the travel time between two customers or between the plant and a customer is equal to the distance between them.

5.2. Results for Small-Sized Instances

The mathematical model is validated by CPLEX. It should be noted that Constraint (13) in the mathematical model is non-linear; therefore, a new variable U_{jh} with $U_{jk} = C_j y_{jk}$ is defined to solve the problem. Then, the following linear Constraints (16) replace Constraint (13).

$$\begin{cases} D_0^h \geq \max_{j \in C} U_{jh}, h = 1, 2, \ldots, H \\ 0 \leq U_{jh} + My_{jh}, j = 1, 2, \ldots, n; h = 1, 2, \ldots, H \\ V_j \leq U_{jh} + M(1 - y_{jh}), j = 1, 2, \ldots, n; h = 1, 2, \ldots, H \end{cases} \quad (16)$$

Table 1 gives the solutions of CPLEX and the improved large neighborhood search (ILNS) algorithm for small-sized cases. The column "Instance" denotes the name of the instance, and "n" and "k" denote the number of customers and the number of vehicles, respectively. The results in Table 1 show that both CPLEX and the ILNS algorithm can get the optimal solution of small-sized test cases, while the running time of the ILNS algorithm is less than that of CPLEX for all of the test cases.

Table 1. Computational results of CPLEX and the improved large neighborhood search (ILNS) algorithm.

	CPLEX		ILNS		
INSTANCE CPLEX	Objective	Time (s)	Objective	Time (s)	Optimal?
Small-n5-k2	302	5	302	3	Yes
Small-n6-k2	349	5	349	3	Yes
Small-n7-k2	397	7	397	3	Yes
Small-n8-k2	720	9	720	3	Yes
Small-n9-k2	1116	210	1116	3	Yes
Small-n10-k2	1225	1381	1225	3	Yes

5.3. Results for Larger-Sized Instances

In order to evaluate the effect of the proposed improved ILNS algorithm for larger-sized instances, we compared it with the initial solution (IS) and a genetic algorithm (GA) [22]. Single machine, multi-customers, and multi-vehicles were considered in [22], while the objective was to minimize the makespan, and an improved genetic algorithm was proposed. For a fair comparison, the running time of GA was set to be equal to the ILNS algorithm for each instance. The ILNS algorithm and GA were coded in C++.

Tables 2–4 provide the results of the instance of Groups A, B, and P, obtained by algorithms IS, ILNS, and GA for the proposed problem. Column "Instance" denotes the instance name. Columns S1, S2, and S3 denote the results obtained by IS, ILNS algorithm, and GA, respectively. Columns T1 and T2 denote the running times of the IS and ILNS algorithms. The running time of GA is equal to the ILNS algorithm. Columns "Gap1" and "Gap2" denote the percentage gap between the IS and ILNS algorithm results, and the GA and ILNS algorithm results, respectively. They are calculated by $Gap1 = 100\% \times \frac{(S1-S2)}{S1}$ and $Gap2 = 100\% \times \frac{(S3-S2)}{S3}$.

Figure 5 provides the tendency in the results of IS, the ILNS algorithm, and GA for instances of Groups A, B and P. It shows that the results of the ILNS algorithm are better than those of IS and GA. Tables 2–4 show that the results of the ILNS algorithm are substantially improved based on IS. The average gaps between the IS and ILNS algorithms are 26.20%, 26.04%, and 22.94% for the three instance groups, respectively. At the same time, the computation time of the ILNS algorithm is short. The average running time of the IS is within 0.1 s, and the average running time of the ILNS algorithm is approximately 50 s. This indicates that the performance of the proposed ILNS algorithm is effective and efficient for not only random instances, but also cluster instances and related instances. It also suggests that the removal heuristics and insertion heuristics work well.

Table 2. Computational results of four algorithms for group A. IS—initial solution; GA—genetic algorithm.

INSTANCE	IS S1	T1 (s)	ILNS S2	T2 (s)	GA S3	Gap1 (S1-S2)/S1 × 100%	Gap2 (S3-S2)/S3 × 100%
A-n32-k5	39,950	0.1	26,793	21	30,560	32.93%	12.33%
A-n33-k5	36,365	0.1	27,038	27	27,104	25.65%	0.24%
A-n33-k6	38,089	0.1	27,500	25	28,257	27.80%	2.68%
A-n34-k5	39,784	0.1	27,097	26	27,953	31.89%	3.06%
A-n36-k5	41,233	0.1	28,538	30	29,629	30.79%	3.68%
A-n37-k5	49,801	0.1	27,846	30	31,928	44.09%	12.79%
A-n37-k6	45,931	0.1	37,024	33	37,389	19.39%	0.98%
A-n38-k5	51,750	0.1	34,040	33	35,199	34.22%	3.29%
A-n39-k5	48,215	0.1	36,046	35	39,128	25.24%	7.88%
A-n39-k6	47,778	0.1	33,226	36	35,404	30.46%	6.15%
A-n44-k6	63,887	0.1	46,951	40	47,775	26.51%	1.72%
A-n45-k6	67,183	0.1	52,041	40	56,139	22.54%	7.30%
A-n45-k7	60,960	0.1	44,569	41	45,427	26.89%	1.89%
A-n46-k7	65,123	0.1	44,634	51	47,779	31.46%	6.58%
A-n48-k7	72,441	0.1	50,220	52	54,918	30.67%	8.55%
A-n53-k7	79,839	0.1	57,642	52	66,360	27.80%	13.14%
A-n54-k7	81,834	0.1	65,353	53	65,671	20.14%	0.48%
A-n55-k9	88,571	0.1	67,390	56	79,548	23.91%	15.28%
A-n60-k9	105,298	0.1	77,041	56	93,972	26.84%	18.02%
A-n61-k9	107,332	0.1	95,840	58	96,831	10.71%	1.02%
A-n62-k8	112,022	0.1	79,393	58	92,145	29.13%	13.84%
A-n63-k9	126,931	0.1	103,647	58	111,543	18.34%	7.08%
A-n63-k10	109,440	0.1	82,116	60	97,488	24.97%	15.77%
A-n64-k9	106,569	0.1	83,404	60	95,853	21.74%	12.99%
A-n65-k9	115,155	0.1	99,977	65	102,061	13.18%	2.04%
A-n69-k9	112,302	0.1	88,760	68	96,742	20.96%	8.25%
A-n80-k10	177,768	0.1	126,095	70	136,001	29.07%	7.28%
Average	77,465	0.1	58,156	46	63,289	26.20%	7.20%

Table 3. Computational results of four algorithms for Group B.

INSTANCE	IS S1	T1 (s)	ILNS S2	T2 (s)	GA S3	GA (S1-S2)/S1 × 100%	Gap2 (S3-S2)/S3 × 100%
B-n31-k5	32,037	0.1	23,086	20	27,400	27.94%	15.74%
B-n34-k5	34,005	0.1	26,357	21	27,840	22.49%	5.33%
B-n35-k5	49,824	0.1	34,661	21	38,894	30.43%	10.88%
B-n38-k6	45,343	0.1	30,791	22	33,623	32.09%	8.42%
B-n39-k5	43,406	0.1	28,290	24	35,428	34.82%	20.15%
B-n41-k6	56,097	0.1	41,899	24	48,512	25.31%	13.63%
B-n43-k6	49,010	0.1	36,450	25	42,848	25.63%	14.93%
B-n44-k7	59,878	0.1	42,445	27	51,955	29.11%	18.30%
B-n45-k5	53,701	0.1	43,150	24	44,859	19.65%	3.81%
B-n45-k6	58,183	0.1	50,010	24	52,044	14.05%	3.91%
B-n50-k7	59,880	0.1	41,784	28	42,563	30.22%	1.83%
B-n50-k8	75,327	0.1	58,666	31	66,615	22.12%	11.93%
B-n51-k7	84,392	0.1	61,684	32	69,275	26.91%	10.96%
B-n52-k7	76,791	0.1	52,150	34	63,476	32.09%	17.84%
B-n56-k7	66,308	0.1	51,942	31	61,185	21.67%	15.11%
B-n57-k7	90,875	0.1	69,093	35	76,394	23.97%	9.56%
B-n57-k9	93,921	0.1	69,093	43	85,416	26.43%	19.11%
B-n63-k10	121,647	0.1	68,553	48	85,632	43.65%	19.94%
B-n64-k9	102,957	0.1	87,335	53	91,611	15.17%	4.67%
B-n66-k9	105,353	0.1	90,749	51	95,558	13.86%	5.03%
B-n67-k10	113,652	0.1	82,307	52	93,042	27.58%	11.54%
B-n68-k9	110,826	0.1	82,378	61	92,653	25.67%	11.09%
B-n78-k10	134,927	0.1	97,012	66	100,653	28.10%	3.62%
AVERAGE	74,710	0.1	55,212	35	62,064	26.04%	11.19%

Table 4. Computational results of four algorithms for Group C.

INSTANCE	IS S1	IS T1 (s)	ILNS S1	ILNS T2 (s)	GA S3	Gap1 (S1-S2)/S1 × 100%	Gap2 (S3-S2)/S3 × 100%
P-n16-k8	7375	0.1	5814	2	6673	21.17%	12.87%
P-n19-k2	15,717	0.1	13,190	2	13,679	16.08%	3.57%
P-n20-k2	17,867	0.1	13,616	2	14,814	23.79%	8.09%
P-n21-k2	16,059	0.1	13,420	3	14,113	16.43%	4.91%
P-n22-k2	16,968	0.1	13,272	3	14,427	21.78%	8.01%
P-n40-k5	52,858	0.1	36,417	26	44,662	31.10%	18.46%
P-n45-k5	63,627	0.1	46,808	25	55,816	26.43%	16.14%
P-n50-k7	93,494	0.1	71,981	31	86,726	23.01%	17.00%
P-n50-k8	95,320	0.1	87,236	34	89,506	8.48%	2.54%
P-n50-k10	83,723	0.1	65,187	45	79,185	22.14%	17.68%
P-n51-k10	68,048	0.1	58,944	47	62,206	13.38%	5.24%
P-n55-k7	119,836	0.1	95,366	52	100,486	20.42%	5.10%
P-n55-k8	108,155	0.1	78,880	56	97,682	27.07%	19.25%
P-n55-k10	105,705	0.1	79,318	56	97,313	24.96%	18.49%
P-n60-k10	114,276	0.1	87,847	51	102,735	23.13%	14.49%
P-n60-k15	107,808	0.1	82,345	52	95,938	23.62%	14.17%
P-n65-k10	194,119	0.1	111,145	55	128,938	42.74%	13.80%
P-n70-k10	174,502	0.1	143,062	55	151,189	18.02%	5.38%
P-n76-k4	227,017	0.1	167,543	60	182,731	26.20%	8.31%
P-n76-k5	228,876	0.1	168,409	62	192,277	26.42%	12.41%
P-n101-k4	334,428	0.1	249,633	91	258,090	25.36%	3.28%
AVERAGE	106,942	0.1	80,449	39	89,961	22.94%	10.91%

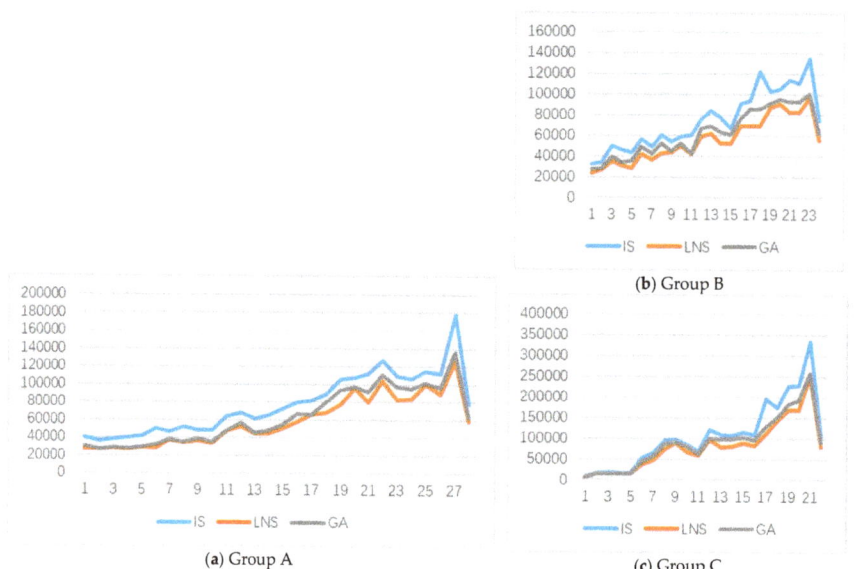

(a) Group A   (b) Group B   (c) Group C

Figure 5. The tendency in the results of three algorithms (IS, ILNS, and GA).

Moreover, the results of the ILNS algorithm are better than those of the GA. For the three instance groups, the average gaps between the ILNS algorithm and the GA are 7.20%, 11.19%, and 10.91%, respectively. The efficiency of the ILNS algorithm could be explained as follows. For the GA in [22], a method "MUS" which adopts some pre-designed rules to change the order of the customers in a route is defined as a local search algorithm. In their transportation phase, the goal is to minimize the maximum delivery time of the orders of each vehicle, "MUS" is employed to find more shortened routing for each vehicle. However, the objective of this paper is to minimize the sum of the order weighted delivery time. In the transportation phase, the goal is to find a method to minimize the sum time of all of the customers receiving services. Obviously, the local search "MUS" is not suitable for solving this goal. For example, reversing all of the points in a TSP route does not affect the solution of the TSP (traveling salesman problem), but reversing all points in a path of this problem would change the solution. Thus, the local search "MI" applied in this paper is more effective than the "MUS" method for minimizing the sum of the order weighted delivery time.

6. Conclusions

This paper explores an integrated production and transportation scheduling problem for perishable products, which is an NP hard problem, aimed at minimizing the total of the order weight delivery time to improve customer service. In the production stage, a single machine is considered, and the order batching and the production sequence of the orders are determined. In the transportation stage, multiple vehicles and multiple customers are considered, and decisions on vehicle routing are made. An integrated mathematical model is built, and the validity is measured by the linear programming software CPLEX, by solving the small-size instances. An improved large neighborhood search (ILNS) algorithm is proposed to solve the larger-size instances. Firstly, a two-stage algorithm constructs an initial solution. The saving algorithm is developed to determine the vehicle routing, and then the optimal production sequence is decided by a certain rule, according to the given vehicle routing. Secondly, several removal and insertion heuristics are designed to destroy and repair the current solution, to generate extensive neighbor solutions to enlarge the solution space. Then, a local optimization algorithm is used to improve the quality of the generated neighbor solutions, which generates more chances to find the optimal solution. Finally, the acceptance rule of the simulated annealing algorithm is used to determine whether to accept the neighbor solution as the new current solution. To validate and evaluate the effectiveness of the proposed ILNS algorithm, the solutions are compared with the corresponding results obtained by the initial solution and the existing genetic algorithm in the literature. The computational results show that the proposed ILNS algorithm substantially improves the initial solution and is more effective than the genetic algorithm.

In the future, we will explore solving the perishable products' integrated production and distribution scheduling problem with exact algorithms, and explore the upper or lower bounds. Moreover, from a practical perspective, considering parallel machines and heterogeneous vehicles with time window constraints are also worthy of being addressed as well.

Author Contributions: Formal analysis, L.L.; Methodology, S.L. All authors have read and agreed to the published version of the manuscript.

Funding: This work was supported by the National Natural Science Foundation of China (no. 71862034, no. 71862035, and no. 71502159), the Scientific Research Funding of Yunnan Department of Education (no. 2017ZZX004), and the Basic Research Foundation of Yunnan Province (no. 2019FB085).

Conflicts of Interest: The authors declare no conflict of interest.

References

1. Amorim, P.; Meyr, H.; Almeder, C.; Almada-Lobo, B. Managing perishability in production-distribution planning: A discussion and review. *Flex. Serv. Manuf. J.* **2013**, *25*, 389–413. [CrossRef]
2. Karaoğlan, İ.; Kesen, S.E. The coordinated production and transportation scheduling problem with a time-sensitive product: A branch-and-cut algorithm. *Int. J. Prod. Econ.* **2017**, *55*, 22.
3. Bashiri, M.; Badri, H.; Talebi, J. A new approach to tactical and strategic planning in production–distribution networks. *Appl. Math. Model.* **2012**, *36*, 1703–1717. [CrossRef]
4. Pundoor, G.; Chen, Z.L. Scheduling a production-distribution system to optimize the tradeoff between delivery tardiness and total distribution cost. *Naval Res. Logist.* **2005**, *52*, 571–589. [CrossRef]
5. Chen, Z.L.; Pundoor, G. Order assignment and scheduling in a supply chain. *Oper. Res.* **2006**, *54*, 555–572. [CrossRef]

6. Li, K.P.; Sivakumar, A.I.; Ganesan, V.K. Complexities and algorithms for synchronized scheduling of parallel machine assembly and air transportation in consumer electronics supply chain. *Eur. J. Oper. Res.* **2008**, *187*, 442–455. [CrossRef]
7. Chen, Z.L.; Vairaktarakis, G.L. Integrated scheduling of production and distribution operations. *Manag. Sci.* **2005**, *51*, 614–628. [CrossRef]
8. Sel, C.; Bilgen, B. Hybrid simulation and mip based heuristic algorithm for the production and distribution planning in the soft drink industry. *J. Manuf. Syst.* **2014**, *33*, 385–399. [CrossRef]
9. Russell, R.; Chiang, W.C.; Zepeda, D. Integrating multi-product production and distribution in newspaper logistics. *Comput. Oper. Res.* **2008**, *35*, 1576–1588. [CrossRef]
10. Chiang, W.C.; Russell, R.; Xu, X.J.; Zepeda, D. A simulation/metaheuristic approach to newspaper production and distribution supply chain problems. *Int. J. Prod. Econ.* **2009**, *121*, 752–767. [CrossRef]
11. Russell, R. A constraint programming approach to designing a newspaper distribution system. *Int. J. Prod. Econ.* **2013**, *145*, 132–138. [CrossRef]
12. Liu, R.; Yuan, B.; Jiang, Z. Mathematical model and exact algorithm for the home care worker scheduling and routing problem with lunch break requirements. *Int. J. Prod. Res.* **2017**, *55*, 558–575. [CrossRef]
13. Mohammadi, S.; Al-e-Hashem, S.M.J.; Rekik, Y. An integrated production scheduling and delivery route planning with multi-purpose machines: A case study from a furniture manufacturing company. *Int. J. Prod. Econ.* **2020**, *219*, 347–359. [CrossRef]
14. Armstrong, R.; Gao, S.; Lei, L. A zero-inventory production and distribution problem with a fixed customer sequence. *Ann. Oper. Res.* **2008**, *159*, 395–414. [CrossRef]
15. Geismar, H.N.; Laporte, G.; Lei, L.; Sriskandarajah, C. The integrated production and transportation scheduling problem for a product with a short lifespan. *J. Comput.* **2008**, *20*, 21–33. [CrossRef]
16. Chen, H.K.; Hsueh, C.F.; Chang, M.S. Production scheduling and vehicle routing with time windows for perishable food products. *Comput. Oper. Res.* **2009**, *36*, 2311–2319. [CrossRef]
17. Viergutz, C.; Knust, S. Integrated production and distribution scheduling with lifespan constraints. *Ann. Oper. Res.* **2014**, *213*, 293–318. [CrossRef]
18. Belo-Filho, M.A.F.; Amorim, P.; Almada-Lobo, B. An adaptive large neighbourhood search for the operational integrated production and distribution problem of perishable products. *Int. J. Prod. Res.* **2015**, *53*, 1–19. [CrossRef]
19. Devapriya, P.; Ferrell, W.; Geismar, N. Integrated Production and Distribution Scheduling with a Perishable Product. *Eur. J. Oper. Res.* **2017**, *259*, 906–916. [CrossRef]
20. Lacomme, P.; Moukrim, A.; Quilliot, A.; Vinot, M. Supply chain optimisation with both production and transportation integration: Multiple vehicles for a single perishable product. *Int. J. Prod. Res.* **2018**, *56*, 4313–4336. [CrossRef]
21. Chen, Z.L. Integrated production and outbound distribution scheduling: Review and extensions. *Oper. Res.* **2010**, *58*, 130–148. [CrossRef]
22. Zou, X.; Liu, L.; Li, K.; Li, W. A coordinated algorithm for integrated production scheduling and vehicle routing problem. *Int. J. Prod. Res.* **2018**, *56*, 5005–5024. [CrossRef]
23. Yan, C.; Banerjee, A.; Yang, L. An integrated production–distribution model for a deteriorating inventory item. *Int. J. Prod. Econ.* **2011**, *133*, 228–232. [CrossRef]
24. Garcia, J.M.; Lozano, S.; Canca, D. Coordinated scheduling of production and delivery from multiple plants. *Robot. Comput. Integr. Manuf.* **2004**, *20*, 191–198. [CrossRef]
25. Garcia, J.M.; Lozano, S. Production and delivery scheduling problem with time windows. *Comput. Ind. Eng.* **2005**, *48*, 733–742. [CrossRef]
26. Asbach, L.; Dorndorf, U.; Pesch, E. Analysis, modeling and solution of the concrete delivery problem. *Eur. J. Oper. Res.* **2009**, *193*, 820–835. [CrossRef]
27. Schmid, V.; Doerner, K.F.; Hartl, R.F.; Stoecher, S.W. A hybrid solution approach for ready-mixed concrete delivery. *Transp. Sci.* **2009**, *43*, 70–85. [CrossRef]
28. Schmid, V.; Doerner, K.F.; Hartl, R.F.; Salazar-González, J.J. Hybridization of very large neighborhood search for ready-mixed concrete delivery problems. *Comput. Oper. Res.* **2010**, *37*, 559–574. [CrossRef]
29. Huo, Y.; Leung, Y.T.; Wang, X. Integrated production and delivery scheduling with disjoint windows. *Discret. Appl. Math.* **2009**, *158*, 921–931. [CrossRef]

30. Geismar, H.N.; Dawande, M.; Sriskandarajah, C. Pool-point distribution of zero-inventory products. *Prod. Oper. Manag.* **2011**, *20*, 737–753. [CrossRef]
31. Amorim, P.; Günther, H.O.; Almada-Lobo, B. Multi-objective integrated production and distribution planning of perishable products. *Int. J. Prod. Econ.* **2012**, *138*, 89–101. [CrossRef]
32. Farahani, P.; Grunow, M.; Günther, H.-O. Integrated production and distribution planning for perishable food products. *Flex. Serv. Manuf. J.* **2012**, *24*, 28–51. [CrossRef]
33. Kopanos, G.M.; Puigjaner, L.; Georgiadis, M.C. Simultaneous production and logistics operations planning in semicontinuous food industries. *Omega* **2012**, *40*, 634–650. [CrossRef]
34. Lee, J.; Kim, B.I.; Johnson, A.L.; Lee, K. The nuclear medicine production and delivery problem. *Eur. J. Oper. Res.* **2014**, *236*, 461–472. [CrossRef]
35. Geismar, H.N.; Murthy, N.M. Balancing Production and Distribution in Paper Manufacturing. *Prod. Oper. Manag.* **2015**, *24*, 1164–1178. [CrossRef]
36. Kergosien, Y.; Gendreau, M.; Billaut, J.C. A Benders decomposition-based heuristic for a production and outbound distribution scheduling problem with strict delivery constraints. *Eur. J. Oper. Res.* **2017**, *262*, 287–298. [CrossRef]
37. Neves-Moreira, F.; Almada-Lobo, B.; Cordeau, J.F.; Guimarães, L. Solving a large multi-product production-Routing problem with delivery time windows. *Omega* **2019**, *86*, 154–172. [CrossRef]
38. Gharaei, A.; Jolai, F. A Pareto approach for the multi-factory supply chain scheduling and distribution problem. *Oper. Res.* **2019**, 1–32. [CrossRef]
39. Sawik, B.; Faulin, J.; Pérez-Bernabeu, E. Multi-Criteria Optimization for Fleet Size with Environmental Aspects. *Transp. Res. Procedia* **2017**, *27*, 61–68. [CrossRef]
40. Kizys, R.; Juan, A.A.; Sawik, B.; Calvet, L. A Biased-Randomized Iterated Local Search Algorithm for Rich Portfolio Optimization. *Appl. Sci.* **2019**, *9*, 3509. [CrossRef]
41. Jiang, T.H. Cat swarm optimization for solving flexible job shop scheduling problem. Computer Engineering and Applications. *Comput. Eng. Appl.* **2018**, *54*, 259–270.
42. Rossi, F.L.; Nagano, M.S. Heuristics for the mixed no-idle flowshop with sequence-dependent setup times. *J. Oper. Res. Soc.* **2019**, 1–27. [CrossRef]
43. Caceres-Cruz, J.; Arias, O.; Guimarans, D.; Riera, D.; Angel, A.J. Rich vehicle routing problem: Survey. *ACM Comput. Surv.* **2015**, *42*, 1–28. [CrossRef]
44. Liu, L.; Li, K.; Liu, Z. A capacitated vehicle routing problem with order available time in e-commerce industry. *Eng. Optim.* **2017**, *49*, 449–465. [CrossRef]
45. Salehipour, A.; Sörensen, K.; Goos, P.; Bräysy, O. Efficient GRASP+VND and GRASP+VNS metaheuristics for the traveling repairman problem. *4OR* **2011**, *9*, 189–209. [CrossRef]
46. Liu, L.; Li, W.L.; Li, K.P.; Zou, X.X. A coordinated production and transportation scheduling problem with minimum sum of order delivery times. *J. Heuristics* **2019**, 1–26. [CrossRef]
47. Ribeiro, G.M.; Laporte, G. An adaptive large neighborhood search heuristic for the cumulative capacitated vehicle routing problem. *Comput. Oper. Res.* **2012**, *39*, 728–735. [CrossRef]
48. Smith, S.L.; Imeson, F. Glns: An effective large neighborhood search heuristic for the generalized traveling salesman problem. *Comput. Oper. Res.* **2017**, *87*, 1–19. [CrossRef]
49. Grimault, A.; Bostel, N.; Lehuédé, F. An adaptive large neighborhood search for the full truckload pickup and delivery problem with resource synchronization. *Comput. Oper. Res.* **2017**, *88*, 1–14. [CrossRef]
50. Rifai, A.P.; Nguyen, H.T.; Dawal, S.Z.M. Multi-objective adaptive large neighborhood search for distributed reentrant permutation flow shop scheduling. *Appl. Soft Comput.* **2016**, *40*, 42–57. [CrossRef]
51. He, L.; de Weerdt, M.; Yorke-Smith, N. Time/sequence-dependent scheduling: The design and evaluation of a general purpose tabu-based adaptive large neighbourhood search algorithm. *J. Intell. Manuf.* **2019**, 1–28. [CrossRef]
52. Eskandarpour, M.; Dejax, P.; Péton, O. A large neighborhood search heuristic for supply chain network design. *Comput. Oper. Res.* **2017**, *80*, 23–37. [CrossRef]
53. He, L.; Liu, X.L.; Laporte, G.; Chen, Y.W.; Chen, Y.G. An improved adaptive large neighborhood search algorithm for multiple agile satellites scheduling. *Comput. Oper. Res.* **2018**, *100*, 12–25. [CrossRef]
54. Haddadi, S.; Cheraitia, M. Iterated local and very-large-scale neighborhood search for a novel uncapacitated exam scheduling model. *Int. J. Manag. Sci. Eng. Manag.* **2018**, *13*, 286–294. [CrossRef]

55. Clarke, G.; Wright, J.W. Scheduling of vehicles from a central depot to a number of delivery points. *Oper. Res.* **1964**, *12*, 568–581. [CrossRef]
56. Ropke, S.; Pisinger, D. A unified heuristic for a large class of vehicle routing problems with backhauls. *Eur. J. Oper. Res.* **2006**, *171*, 750–775. [CrossRef]
57. Kurz, M.E.; Askin, R.G. Heuristic scheduling of parallel machines with sequence-dependent set-up times. *Int. J. Prod. Res.* **2001**, *39*, 23. [CrossRef]
58. Hemmelmayr, V.C. Sequential and parallel large neighborhood search algorithms for the periodic location routing problem. *Eur. J. Oper. Res.* **2015**, *243*, 52–60. [CrossRef]

© 2020 by the authors. Licensee MDPI, Basel, Switzerland. This article is an open access article distributed under the terms and conditions of the Creative Commons Attribution (CC BY) license (http://creativecommons.org/licenses/by/4.0/).

MDPI
St. Alban-Anlage 66
4052 Basel
Switzerland
Tel. +41 61 683 77 34
Fax +41 61 302 89 18
www.mdpi.com

Mathematics Editorial Office
E-mail: mathematics@mdpi.com
www.mdpi.com/journal/mathematics

www.ingramcontent.com/pod-product-compliance
Lightning Source LLC
LaVergne TN
LVHW070651100526
838202LV00013B/933